ANCIENT

The Lost Art of the Anglo-Saxon World

The Sacred and Secular Power of Embroidery

Alexandra Lester-Makin

OXBOW | books
Oxford & Philadelphia

First published in 2019. Reprinted in the United Kingdom in 2021 by
OXBOW BOOKS
The Old Music Hall, 106–108 Cowley Road, Oxford OX4 1JE

and in the United States by
OXBOW BOOKS
1950 Lawrence Road, Havertown, PA 19083

© Oxbow Books and the author 2019

Paperback Edition: ISBN 978-1-78925-144-9
Digital Edition: ISBN 978-1-78925-145-6 (ePub)

A CIP record for this book is available from the British Library

Library of Congress Control Number: 2019948164

All rights reserved. No part of this book may be reproduced or transmitted in any form or by any means, electronic or mechanical including photocopying, recording or by any information storage and retrieval system, without permission from the publisher in writing.

Printed in the United Kingdom by Short Run Press

Typeset in India for Casemate Publishing Services. www.casematepublishingservices.com

For a complete list of Oxbow titles, please contact:

UNITED KINGDOM	UNITED STATES OF AMERICA
Oxbow Books	Oxbow Books
Telephone (01865) 241249	Telephone (610) 853-9131, Fax (610) 853-9146
Email: oxbow@oxbowbooks.com	Email: queries@casemateacademic.com
www.oxbowbooks.com	www.casemateacademic.com/oxbow

Oxbow Books is part of the Casemate Group

Front cover: Maaseik, detail of roundel strip 1, with thanks to the Royal Institute for Cultural Heritage, Brussels.
Back cover: Reel of gold thread from Southampton, image: Elaine Wakefield, © Wessex Archaeology.

To my family, Ian and Edward

Contents

Acknowledgements vi
List of illustrations vii

1. Introducing Anglo-Saxon embroidery 1
2. The data and the difficulties 31
3. Kempston: the biography of an embroidery 57
4. Embroidery and Anglo-Saxon society 77
5. Early medieval embroidery production in the British Isles 101
6. Conclusion: embroidery in context 141

Appendix 1: Catalogue 149
Appendix 2: Glossary of terminology 195
Appendix 3: 211
    Table 1A–G: Table of surviving insular embroideries 212
    Table 2: Loose gold thread 224
    Table 3: Pieces with no surviving thread 225
Appendix 4: Chronological table 229
Bibliography 231

Acknowledgments

There are a large number of people I would like to thank; firstly, my husband Ian and my family for their continued support throughout the process. Secondly, those who have given me advice, guidance and help (in alphabetical order): Spencer Bailey, Martin Biddle, Sue Brunning, Elizabeth Coatsworth, Melanie Giles, Ann Zanette Tisgaridas Glørstad, Catherine Hills, Sylvette Lemagnen and the staff at the Musée de la Tapisserie de Bayeux, Frances Lennard, Claire Marsland, Christopher Monk, Louise Mumford, Anja Neskens and staff at the Maaseik Museums Department and the Royal Institute for Cultural Heritage in Belgium, Gale Owen-Crocker, Michael Pinder, Frances Pritchard, the staff at the Basilica Ambrosiana in Milan, the Durham University Conservation Centre, English Heritage Archive, the Museum of London Archive, Oxbow Books, York Archaeology Trust and Yorkshire Museum.

Thirdly, I would like to thank those who peered reviewed the book manuscript: Susanna Harris, Lynn Hulse and Frances Pritchard.

Finally, I would like to thank Anna Henderson for her help and guidance with reorganising the research and key ideas from my doctoral thesis (Manchester University, 2017): *Embroidery and its Contexts in the British Isles and Ireland during the Early Medieval Period* (AD 450–1100), and with editing the resulting book.

Publication of the coloured plates was made possible by a grant from the Scouloudi Foundation in association with the Institute of Historical Research.

The author would like to thank the C&TA for the Geoffrey Squire Memorial Fund grant towards the cost of images.

List of illustrations

Black and white images

Figure 1   Sutton Hoo A: The seam
Figure 2   The Llangorse Textile
Figure 3   York A: Line drawing of the pouch
Figure 4   York B: Line drawing of the embroidered hem
Figure 5   Sutton Hoo B: Line drawing of the embroidery
Figure 6   Alfriston: Two buckles with embroidery attached
Figure 7   Line drawing of the Worthy Park embroidery
Figure 8   Mitchell's Hill wrist clasps (AN1909.487.[i]).
Figure 9   Line drawing of the Winchester shoe
Figure 10 Line drawing of Oseberg A
Figure 11 Line drawing of Oseberg B
Figure 12 Oseberg C
Figure 13 Oseberg E
Figure 14 Oseberg F
Figure 15 Oseberg G
Figure 16 Oseberg H
Figure 17 Oseberg I
Figure 18 Reel of gold thread from Southampton
Figure 19 Diagram 1: Distribution of surviving embroideries by period
Figure 20 Diagram 2: Proportions of different object types
Figure 21 Diagram 3: Condition of surviving embroideries
Figure 22 Diagram 4: Ground fabric materials
Figure 23 Diagram 5: Categories of surviving embroidery threads
Figure 24 Diagram 6: Embroidery stitches utilised on surviving embroideries
Figure 25 Diagram 7: Relative proportions of surviving design elements drawn from the natural world
Figure 26 Diagram 8: Relative proportions of surviving composite and symbolic elements drawn from the cultural environment
Figure 27 Diagram 9: Relative proportions of surviving embroideries showing simple repetitive pattern, or no pattern
Figure 28 Line drawing of the Kempston embroidery
Figure 29 Line drawing of the Kempston embroidery
Figure 30 Embroidery from Mammen
Figure 31 A bracteate die from Castledyke South, Barton-on-Humber, Lincolnshire

Figure 32 Line drawing of the design from the shield strip discovered in mound 1, Sutton Hoo, Suffolk
Figure 33 The embroidery from Røn, Norway
Figure 34 Line drawing of metal button clasps from Øvstebo, Vindafjord, Ro in Norway
Figure 35 Detail of Oseberg G showing the stylised animal
Figure 36 Detail of tablet woven band 34D from Oseberg, showing a swastika cross
Figure 37 Friskerton sword, Lincolnshire
Figure 38 Gilling Beck sword, North Yorkshire
Figure 39 Fetter Lane sword hilt, London
Figure 40 Detail of AN1890.14 the Abingdon Sword hilt
Figure 41 Colyton cross shaft, Devon
Figure 42 Detail of the central motif showing a bird inhabiting a vine scroll: Colyton cross shaft, Devon
Figure 43 The 7th-century gold and garnet *cicade* from Horncastle, Lincolnshire
Figure 44 The upper cross shaft from Todber, Devon
Figure 45 Line drawing of an embroiderer from Exodus XXXVI, 8–19, *The Byzantine Octateuch*
Figure 46 The author at her embroidery frame, using a slate frame resting on trestles
Figure 47 The author at her embroidery frame, both hands are positioned above the frame as she threads a needle
Figure 48 Detail of women spinning and weaving, *Utrecht Psalter*
Figure 49 Line drawing of structures 4886 and 8148, Brandon, Suffolk
Figure 50 Line drawing of the Ingleby fragments as they would have originally been joined together
Figure 51 Looped stitch, experimental piece by author
Figure 52 London A: detail of the embroidered top band, S-plied sewing thread, front view
Figure 53 London A: detail showing S-plied sewing thread
Figure 54 London A: detail showing the folded leather being held in place by the top line of stitching
Figure 55 A strap-end from York, sf 7306, decorated with an entwinned knot and animal features
Figure 56 Oseberg J
Figure 57 Oseberg J
Figure 58 Back stitch
Figure 59 Binding stitch
Figure 60 Broken diamond twill weave
Figure 61 Buttonhole/blanket stitch
Figure 62 Chain stitch: a) single line; b) two rows
Figure 63 Chevron twill weave

Figure 64 a) Couched metal threads; b) couched work, sometimes called tracey stitch
Figure 65 Diamond twill weave
Figure 66 Herringbone stitch
Figure 67 Laid-work
Figure 68 Looped stitch: a) Dublin, Sutton Hoo A, York; b) Orkney Hood; c) Ingleby
Figure 69 Overcast
Figure 70 Over sew
Figure 71 Plait
Figure 72 a) Plait stitch; b) raised plait stitch; c) a view of plait stitch and raised plait stitch when complete
Figure 73 a) S-ply; b) Z-ply; c) plied thread
Figure 74 Running stitch
Figure 75 Satin stitch
Figure 76 Seed stitch
Figure 77 Slip hem
Figure 78 Soumak: a) thread woven through the ground fabric; b) compacted thread
Figure 79 a) S-spun; b) Z-spun
Figure 80 Split stitch
Figure 81 a) Stem stitch; b) stem stitch, two rows
Figure 82 Tabby weave
Figure 83 a) Tunnel stitch; b) tunnel stitch, front
Figure 84 Twill weave a) 2×2; b) 2×1
Figure 85 Underside couching: a) top view; b) side view; c) reverse view
Figure 86 Amice and Alb
Figure 87 Stole: as worn by a priest
Figure 88 Dalmatic
Figure 89 Chasuble (*casula*) and Orphrey
Figure 90 Pallum and Maniple
Figure 91 Stole: as worn by a deacon and bishop
Figure 92 Boot
Figure 93 Shoe
Figure 94 a) Normal sole; b) triangular sole

Colour plates

Plate 1 a) The Orkney Hood; b) Worcester: The embroidered fragments in their original frame
Plate 2 a) Sutton Hoo A: Textile with seam covered in embroidery to the left; b) Kempston embroidery
Plate 3 a) Durham D: Stole; b) Durham E: Maniple

Plate 4   Durham C: Embroidered band known as 'Maniple II' or a girdle; a) front view; b) reverse view

Plate 5   a) Bayeux Tapestry: Scene 30, 'Here Harold sits as King of the English'; b) London A: shoe with embroidered top band

Plate 6   a) The eight Maaseik embroideries; b) Coventry: Child's leather turnshoe, right foot, with embroidered vamp

Plate 7   a–b) Ingleby: Carbonised metal thread embroidery: a) top, front view; b) bottom, front view; c) Durham D: Fragment of stole from Ushaw College

Plate 8   a) Durham A: Tablet-woven soumak band with possible embroidery; b) Durham F: Fabric with embroidery from the grave of William of St Calais

Plate 9   a) Durham G: Fabric with embroidery from the grave of William of St Calais; b) Milan: Gold work embroidery in its 1940s frame

Plate 10   The composite textile or *casula* of Sts Harlindis and Relindis, with the eight embroideries *in situ*

Plate 11   Oseberg J: Colour drawing of section 12B5

Plate 12   a) Milan: Detail of gold embroidery attached to the original silk ground fabric; b) Durham A: Detail of faded coloured threads; c) Durham A: Detail of possible embroidered wrapping stitch; d) Kempston: copper alloy box in which the embroidery was found

Plate 13   Microscopic images of the Kempston embroidery; a) chain stitch circled. The arrow to the right points to the possible split stitch; b) line of stem stitch circled; c) the two sections placed next to each other; d) the break in stitching and the small individual stitches

Plate 14   Kempston: a) line drawing and b) coloured drawing of embroidery design

Plate 15   a) Great gold buckle, burial mound 1, the ship burial, Sutton Hoo; b) shoulder clasps, Taplow

Plate 16   Book of Durrow carpet page

Plate 17   a) Kingston broach: detail; b) Sword hilt collar from the Staffordshire Hoard

Plate 18   a) Durham E: maniple, detail, halo of the deacon Peter; b) line drawing, braid 9 from Durham; c) Durham E: maniple, detail, halo of Pope Gregory; d) line drawing, braid 10 from Durham; e) Durham D: Stole both end tabs, reverse

Plate 19   Maaseik: a) roundel strip detail; b) detail of a surviving pearl; c) roundel strip 2

Plate 20   a) Books of Kells: moths highlighted; b) Milan: detail of couched gold thread; c) Durham D: stole, detail of couched gold thread; d) Maaseik B: detail of couched gold thread

Plate 21   Milan a) second square from left: duck with head turned back, third square from left: stylised insect; b) detail of gold work design and layout

Plate 22   Dedication page of Bede's *Life of St Cuthbert*

Plate 23   Maaseik A: arcade strip detail

List of illustrations

Plate 24 Maaseik A: a) detail, outlines worked first then reworked after the filling is complete (highlighted by arrow); b) detail, stitching following triangular motif shapes; c) detail, more randomly orientated stitching; d) detail, stitching follows line of curve and colours are worked in blocks

Plate 25 Maaseik B a) detail, stitches are worked in one direction; b) detail, combination of stitched lines and chevrons; c) detail, couching stitches worked in rows (circled); d) detail, couched gold thread

Plate 26 a) Maaseik A: detail, couching thread worked double over each gold thread (circled); b) Maaseik B: detail, gold thread bent precisely round corners; c) Maaseik A: detail, filling threads cross outline threads; d) couched gold catches silk infilling threads; e) and f) Worcester: details of underside couching used as a filling stitch, front and reverse

Plate 27 Worcester: a) detail of underside couching used as an outline, front; b) detail of underside couching used as an outline, reverse

Plate 28 Comparative examples of silk work: a) Durham E; b) Worcester; c) orphrey (AD 1310–1325)

Plate 29 Comparative stance between bishops: a) Worcester; b) Bayeux Tapestry

Plate 30 Comparative stance between figures: a) Worcester; b) Bayeux Tapestry

Plate 31 Comparative buildings: a) and c) Worcester; b) and d) Bayeux Tapestry

Plate 32 Maaseik C: two of the four monograms

Chapter 1

Introducing Anglo-Saxon embroidery

This book is about the production and use of embroidery and its relevance in society in the British and Irish archipelagos during the early medieval period, AD 450–1100. It is an interdisciplinary work encompassing study of the design, technique and construction of the embroideries themselves in the archaeological, ecclesiastical and museum contexts in which they are held, and in conjunction with the surviving documentary sources and associated archaeological evidence. By taking an integrated approach and contextualising embroidery within the early medieval period, we can establish its significance as material culture, while at the same time, using it to enhance our understanding of the early medieval world.

Defining the early medieval period

As the medieval archaeologist John Moreland notes, it is not always helpful to compartmentalise eras. Preoccupation with the beginning and end dates of a culture can result in a distorted reading of the evidence, 'objects, institutions, concepts are treated either as precedent or as relic – not as active in the construction of people and society in their own times'.[1] Although the focus of this particular research is the early medieval period in the British and Irish archipelagos and dates have been given as a necessity for confining research parameters, they are intended as a guide to the development of a culture, not in order to isolate embroidery from its own history. Two of the embroideries discussed here, the Orkney hood and the Worcester fragments (remains of what was probably a stole and maniple, worked in silk and silver-gilt

---

1 J. Moreland, *Archaeology, Theory and the Middle Ages* (London: Duckworth, 2010), 8.

threads on a silk ground fabric) (Pl. 1a, b), may date from outside the specified timeframe between *c.* 250 and 615 and the 11th and 12th centuries respectively, demonstrating that embroidery was, and is, a manifestation of continuity in cultural taste and technical knowledge.

The early medieval period in England is defined as the six centuries from the cessation of central Roman imperial rule in AD 410 to the Norman Conquest of Anglo-Saxon England in 1066.[2] Although the people of the territories known today as Wales, Scotland and Ireland did not experience invasion at the hands of the Anglo-Saxons and Normans to the same extent, their history was already entwined with England during this period. Examples include the Scots of Ireland invading and settling in Scotland, and extending their kingdoms into north-west zones of Anglo-Saxon England; the Picts of Scotland raiding the northern Anglo-Saxon kingdoms; and the Welsh incursions into the north-western and western parts of the country. From the 8th century Vikings raided and settled in Ireland as well as eastern England and parts of Scotland, and in the later third of the 11th century the Normans encroached into Wales and Ireland.[3]

Creative influences, including embroidery, move with people. Design, fashion, materials and working techniques would have spread throughout the countries that make up the archipelagos, moving and morphing as they were encountered by different populations. Indeed, this evolution can be seen in other forms of decorative dress accessory such as metalwork.[4] The development and use of embroidery in the different geographical regions is therefore as entwined as the peoples who lived in them. For this reason, embroidery from the regions should be studied as a corpus, not as groups of work that emerged independently of each other. To date, only Elizabeth Coatsworth has treated these works as a single corpus.[5]

My own study begins during a transitional phase within the archipelagos, with the decline of central Roman influence and the ascendency of Germanic peoples within eastern and southern England. There is still much debate and research regarding

---

[2] I have used the term 'central Roman' to denote the Roman Empire as opposed to the native Romanised elite. For discussions of the end of central Roman control and the beginning of Anglo-Saxon influence see for instance: C. Wickham, *The Inheritance of Rome: a history of Europe from 400–1000* (London: Allen Lane, 2009; repr. London: Penguin Books, 2010); B. Cunliffe, *Britain Begins* (Oxford: Oxford University Press, 2013), 401–426; N.J. Higham and M.J. Ryan, eds, *The Anglo-Saxon World* (London: Yale University Press, 2013), 41–111.

[3] B. Yorke, *The Conversion of Britain 600–800* (Harlow: Pearson Education, 2006), 33–61, 269–270; H. Mayr-Harting, *Religion, Politics and Society in Britain 1066–1272* (Harlow: Pearson Education, 2011), 47–49.

[4] For discussions relating to the development of metalwork design associated with dress see: G.R. Owen-Crocker, *Dress in Anglo-Saxon England, revised and enlarged* (Woodbridge: Boydell Press, 2004), 316–320; P. Walton Rogers, *Cloth and Clothing in Early Anglo-Saxon England: AD 450–700* (York: Council for British Archaeology, 2007), 111–138.

[5] E. Coatsworth, 'Stitches in Time: establishing a history of Anglo-Saxon embroidery', in *Medieval Clothing and Textiles,* 1 (2005), 1–27.

this period; but Gildas (d. *c.* 570?) and Bede (*c.* 673–735) are usually taken as sources of documentary evidence for the Germanic tribes' violent over-throw of the native Britons.[6] However, Nicholas Higham and Martin Ryan argue that the archaeological evidence does not confirm this, suggesting that the transition occurred over a much longer time-frame, from the early- to mid-5th century, and with little fighting. It would seem that the Germanic peoples' integration with the local population involved a merging of the two cultures, particularly along the east and southern coasts of modern day England.[7]

Despite the fact that Bede's work uses approximations for dates,[8] a mixture of convention together with supporting archaeological evidence has led scholars to refer to AD 450 as the beginning of the Anglo-Saxon period in England (and of the early medieval era), and for this reason I have used it as the start date for this survey.[9] The rationale for the end date – 1100 – is easier to justify. Although the Norman invasion and conquest of England took place in 1066, scholars now agree that the change was not abrupt.[10] Although many of the elite were killed or replaced and church leaders were supplanted by allies of William of Normandy (the Conqueror) (*c.* 1028–1087), for the rest of society change was more gradual.[11] Daily life may have been interrupted, but the shift in material culture took time to evolve. For instance, Norman appreciation of Anglo-Saxon textiles and embroidery is demonstrated by the fact that William's wife, Matilda (*c.* 1031–1083) commissioned a cope to be made or embroidered by Aldret of Winchester's wife and another robe embroidered 'in England', both of which she bequeathed to the Church of the Holy Trinity in Caen, '*Ego Mathildisregina do Sancte Trinitati Cadomi casulam quam apud Wintoniam operator uxor Aldereti et clamidem operatam ex auro que est in camera mea ad cappam faciendam… ac vestimentum quod operator in Anglia …*'[12] Indeed, Matilda was commissioning people employed in this work prior to 1066.

---

6 Yorke, *Conversion of Britain*, 1, 15–22. For detailed analysis of documentary sources see S. Yeates, *Myth and History: ethnicity and politics in the first millennium British Isles* (Oxford: Oxbow Books, 2012).
7 Higham and Ryan, *Anglo-Saxon World*, 75–111.
8 As the editors of the 1969 edition point out: B. Colgrave and R.A. Mynors, eds, *Bede's Ecclesiastical History of the English People* (Oxford: Clarendon, 1969), i. 15 (49, ftn. 3).
9 In using the term 'Anglo-Saxon' I am aware of the continued debates surrounding the use of ethnic labels, see for example, Higham and Ryan, *Anglo-Saxon World*, 7–10; Yorke, *Conversion of Britain*, 3.
10 See C. Dyer, *Making a Living in the Middle Ages: the people of Britain 850-1520* (London: Yale University, 2009), 71–99 (71); D. Griffiths, 'The Ending of Anglo-Saxon England: identity, allegiance, and nationality', in *The Oxford Handbook of Anglo-Saxon Archaeology*, eds H. Hamerow, D.A. Hinton and S. Crawford (Oxford: Oxford University, 2011), 62–78; Higham and Ryan, *Anglo-Saxon World*, 397–426.
11 Dyer, *Making a Living*, 80–91; Griffiths, 'Ending of Anglo-Saxon England', 63.
12 L. Musset, ed., *Les Actes de Guillaume le Conquérant et de la Reine Mathilde pour les Abbayes Caennaises* (Caen: Société des Antiquaires de Normandie, 1967), 112–113; quotation trans. [Mrs] A.G.I. Christie, *English Medieval Embroidery: a brief survey of English embroidery dating from the beginning of the tenth century until the end of the fourteenth* (Oxford: Clarendon Press, 1938), 32.

A surviving embroidery attests to the continuation of Anglo-Saxon material culture beyond 1066: the Bayeux Tapestry, which scholars believe was made in *c.* 1077 in England.[13] It is in part as a result of this textile's existence that this study concludes after that date, at the end of the reign of William of Normandy's son, William II (*c.* 1056–1100). By then, Norman rule was firmly cemented and Anglo-Saxon material culture would have adapted under Norman influence.

Defining embroidery

Defined at face value, embroidery is simply 'the embellishment of fabrics by means of needle-worked stitches'.[14] However there is more to it than this, because embroidery can function in multiple ways at the same time. It can be used to decorate things and create artistic effects and images while also reinforcing and hemming fabrics, darning or joining them. A good example is the Orkney hood, a child's headdress found in a bog in St Andrew's Parish, Orkney, which has been radiocarbon dated to between AD 250 and 615.[15] The hood, which would have covered the head and shoulders, is made from recycled wool fabric and tablet-woven bands and shows signs of wear in several places. The larger patches have been darned; narrow splits have been joined back together using chain stitch (see Glossary), the edges of the fabric have been whip stitched (see Glossary) and in another area, a tablet-woven band has been attached to the main body of the hood with a complicated variation of looped stitch (see Glossary).[16] This single textile demonstrates functional need being accomplished with three different examples of decorative flair, showing that embroidery requires nuanced analysis that takes into account its multiple roles.[17]

Extant embroideries show that during the early medieval period decoration of textiles took many forms. People used all materials available to them – wool, linen, silk, metal threads and precious stones – to decorate secular clothing, ecclesiastical vestments and soft furnishings. They were innovative in their use of embroidery,

---

13 For an overview of the debate around this see E. Carson Paston and S.D. White with K. Gilbert, *The Bayeux Tapestry and its Contexts: a reassessment* (Woodbridge: Boydell, 2014), 2, 65, 261.
14 I. Emery, *The Primary Structures of Fabrics* (London: Thames and Hudson, 1980; repr. 2009), 233.
15 R.E.M. Hedges, R.A. Housley, C. Bronk-Ramsey and G.J. van Klinken, 'Radiocarbon Dates from the Oxford AMS System: Archaeometry datelist 16', *Archaeometry*, 35 (1993), 147–167 (155); T. Gabra-Sanders, 'The Orkney Hood, Re-Dated and Re-Considered', in *The Roman Textile Industry and its Influence: a birthday tribute to John Peter Wild*, eds P. Walton Rogers, L. Bender Jørgensen and A. Rast-Eicher (Oxford: Oxbow Books, 2001), 99–104 (99).
16 A.S. Henshall, 'Early Textiles Found in Scotland', *Proceedings of the Society of Antiquaries of Scotland*, 86 (1951–2), 1–29 (11, 13–14). For descriptions of all stitches discussed, see Glossary. Modern stitch names have been used throughout.
17 See also A. Morrell, *The Migration of Stitches and the Practice of Stitches as Movement* (Ahmedabad: D.S. Mehta, 2007): Morrell is undertaking the same exercise for modern embroidery stitches. It should also be noted that there is no evidence to suggest whether early medieval embroiderers distinguished between embroidery and needlework.

Figure 1. Sutton Hoo A: the seam (measurements: embroidery: 100 × 12-40 mm), © Trustees of the British Museum

using it to decorate and join seams, and mimic more expensive, rarer fabrics. A fragmented textile from Sutton Hoo (known as 'Sutton Hoo A') (Pl. 2a), and a small piece of wool embroidery from Kempston (Pl. 2b), illustrate these points well. The Sutton Hoo textile dates to the early 7th century and was probably a pillow or bag. It was constructed from pieces of what was possibly a two-tone fabric cut and sewn together. The surviving seam was constructed using a matching thread. The stitch used to join the two pieces of fabric together was transformed from a plain functional sewing stitch into an elaborate looped stitch that covered the seam while sewing it securely (Fig. 1 and see Glossary). The Kempston fragment (the focus of Chapter 3), which also dates to the 7th century, was worked in fine plied wool threads that looked so similar to silk that they were mistaken for just that until the embroidery was expertly examined in the 1970s by Elisabeth Crowfoot.[18]

---

18 E. Crowfoot, 'Textile Fragments from "Relic-Boxes" in Anglo-Saxon Graves', in *Textiles in Northern Archaeology: NESAT III: textile symposium in York, 6-9 May 1987*, eds P. Walton and J.-P. Wild (London: Archetype Publications, 1990), 47–56.

People also appear to have used embroidery to symbolic effect. The use of metal threads in ecclesiastical embroideries is a prime example. The effect of the image adorning the garment was enhanced by the way in which the metal threads were manipulated and patterned by the silk couching threads to create areas of light and dark, shine and dullness. The metal threads and silk fabrics would have created a spectacle as the wearer walked through dimly lit buildings. Light from small windows and internal illumination would have been reflected and fractured by the embroideries. As the textiles moved with the wearer they would dazzle the viewer, creating an ethereal image that had the intention of bringing to mind the spiritual. Moreover, the awe inspired by the expense in production time and material costs would have underpinned and cemented the power and authority of the Church, and those who wore and commissioned the vestments.

In this way, embroidery during this period was not just about the decorative and the functional. It became a tool to create, confirm and strengthen power and authority within society and, therefore, an important aspect of early medieval social and material culture. The study of embroidery not only gives a unique insight into fibre and textile production, development of working practices and artistic and technical skills, and the status of workers, patrons and users (especially women); it also provides insight into early medieval mind-sets.

Nearly all scholars of embroidery history cite a very small number of the more famous early medieval pieces before moving rapidly on to the 12th century. Others start their analyses at *c.* 1150 when the development of *Opus Anglicanum*, a style of embroidery that was celebrated throughout medieval Europe, contributing to the renown of English embroidery-workers, became dominant. The many sumptuous surviving examples of *Opus Anglicanum* housed in churches, cathedrals, museums and private collections across the world helps account for this preoccupation. In contrast there are not as many known examples of embroidery dating to pre-1100, with the Cuthbert embroideries at Durham:[19] three items embroidered with gold and silk on silk ground fabric, consisting of a stole (Pl. 3a), a maniple (Pl. 3b), and an item known as 'maniple II' or 'the small maniple' (Pl. 4a, 4b), which is more likely to have been a pair of ribbons stitched together at some point;[20] and the better known Bayeux Tapestry (Pl. 5a): a long, thin narrative embroidery depicting the Norman Conquest in wool on linen, being the only ones cited in the majority of published works. In recent decades, however, the corpus of surviving embroideries has grown, due largely to fragmented pieces found on excavation sites, the prime examples being from Llangorse in Wales (Fig. 2), and at York (Fig. 3, 4), where two leather- and textile-based embroideries have been discovered, and three similar instances in London (Pl. 5b). At the latter two sites, examples of loose gold thread, which experts have suggested

---

19 Note that two pieces found in the tomb of Bishop William of St Calais are also located at Durham.
20 See below, brief discussion of Coatsworth's work at 14–15.

Figure 2. The Llangorse Textile (textile: 8000 cm², embroidery: c. 780 cm²), © National Museum of Wales

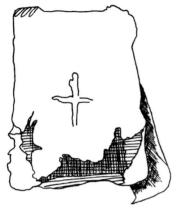
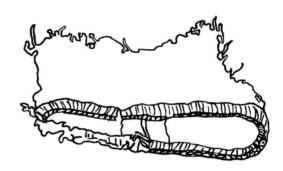

Figure 3. York A: Line drawing of the pouch (pouch: 33 × 25-30 mm, embroidery: 13 × 8 mm), © Alexandra Lester-Makin, after Walton Rogers (1989)

Figure 4. York B: Line drawing of the embroidered hem (textile: 92 × 45 mm, embroidery: 160 × 5 mm),© Alexandra Lester-Makin, after WaltonRogers (2004)

were used for embroidery, and leather and textiles thought to have been embroidered, have also been discovered (see Appendix 3: Table 2 and Table 3).[21] Although these finds have been analysed by archaeological textile experts and written up individually in specialist reports, they are rarely, if ever, included in general surveys of embroidery history. In the light of this expansion of the corpus of material it is more important than ever to study the surviving examples, place them in their social and material contexts and afford them the status they deserve.

The development of embroidery scholarship

The study of the history of medieval embroidery has been largely concerned with *Opus Anglicanum*, gold and silk embroideries produced in England between *c.*1200 and

---

[21] For further information regarding the contexts in which these finds were made, see the following. On London: F. Pritchard, 'Leather Work', in *Aspects of Saxon-Norman London: II Finds and Environmental Evidence*, ed. A. Vince (London: London and Middlesex Archaeological Society, 1991), 211–240; P. MacConnoran with A. Nailer, '8.15 Leather Items', in *The London Guildhall: an archaeological history of a neighbourhood from early medieval to modern times*, II, eds D. Bowsher, T. Dyson, N. Holder and I. Howell, Museum of London Monograph 36 (London: Museum of London Archaeology, 2007), 479–486; A. Nailer and P. Reid with P. MacConnoran, 'Leather and Shoes', in *The Development of Early Medieval and Later Poultry and Cheapside Excavations at 1 Poultry and Vicinity, City of London*, eds M. Burch and P. Treveil with D. Keene (London: Museum of London Archaeology, 2011), 332–341. On York: D.Tweddle, 'Comments on the use of the Pouch as a Reliquary', in *Textiles, Cordage and Raw Fibre from 16-22 Coppergate*, The Archaeology of York 17/5, ed. P. Walton (London: Council for British Archaeology, 1989), 378–338; Q. Mould, I. Carlisle and E. Cameron, eds, *Leather and Leatherworking in Anglo-Scandinavian and Medieval York*, The Archaeology of York, The Small Finds 17/16 (York: Council for British Archaeology, 2003); P. Walton Rogers 'Textiles, Cords, Animal Fibres and Human Hair', in *28-9 High Ousegate, York, UK*, eds N. Macnab and J. McComish (York: York Archaeological Trust, 2018), 14–41.

1400, while less attention has been paid to pre-Conquest 'early medieval' material, that dating to between 450 and 1066. The existing body of research is inconsistent, in part due to the cycle of fashionable research topics, but also because of the complicated cross-curricular nature of the subject. Surviving pieces found in churches or cathedrals fall under the care of conservators and librarians or archivists, while those found within archaeological contexts are studied by textile archaeologists, textile historians, conservators and curators. These specialists may also commission specific scientific analyses of individual embroideries, for example, the identification of fibres, dyes and dates through radiocarbon dating, but normally as part of wider investigation into the object. As a result, publishing on early medieval embroidery falls into a number of distinct categories. There are introductions to the most famous embroideries such as the Cuthbert embroideries, the Bayeux Tapestry and, less frequently, the Maaseik embroideries (Pl. 6a) (see below). The Maaseik embroideries are a set of eight embroideries believed to have originated as decorative bands, probably from secular garments, given to the shrine of the sister saints Harlindis and Relindis at Aldeneik, Belgium); descriptions of individual pieces in exhibition catalogues or encyclopaedias; embroidery used as a means of contextualising other art forms or as part of a wider discussion on textiles; and detailed analysis of individual pieces placed in context in the body, or appendices, of archaeological reports.

It was not until 1938 that a major study of English-produced medieval embroidery by Ada Grace Ida Christie, here referred to as Grace Christie, was published. Prior to that, amateur interest in embroidery came from antiquarians, who subdivided along areas of particular expertise as professional academic fields developed (particularly archaeology, history, art history and textile history), and from women with experience of embroidery as an occupation. Amateur interest coincided with an upsurge in interest in medieval embroidery in Britain as a result of the religious revival of the 19th century, and at a time when ideas of national and imperial identity were being debated. The Gothic Revival Movement, in which the designer Augustus Pugin (d. 1852) was prominent, drew inspiration from a pre-Reformation medieval past after the 1829 Roman Catholic Relief Act liberalised attitudes towards religious practice. The second generation of Gothic Revivalists, including the ecclesiastical antiquarian Dr Daniel Rock (d. 1871), developed this interest as a scholarly subject.[22] British exhibitions such as the 1851 *Great Exhibition of the Works of Industry of all Nations*, brought medieval embroidery to the attention of the wider public, while The Royal School of Art Needlework (est. 1872), the Leek Embroidery Society (est. 1879/1880) and professional needlewomen such as Mrs Anastasia Dolby (d. 1873)

---

22 [Rev.] D. Rock, *South Kensington Museum Textile Fabrics* (London: Chapman and Hall, 1870); [Rev.] D. Rock, *Textile Fabrics* (London: Chapman and Hall, 1876).

revived interest in medieval embroidery techniques.[23] A little later, William Morris and the Arts and Crafts Movement would lead to the establishment of the Central School of Arts and Crafts (1896) where May Morris, Morris's daughter, taught and Grace Christie studied.

The historians William R. Lethaby and Albert F. Kendrick were both part of the network involved in dispensing Arts and Crafts ideas and each developed an interest in medieval embroidery. Like Pugin, Lethaby (1857–1931) trained as an architect and believed study of medieval embroidery could inform other areas of artistic research. He also saw embroidery as sufficiently important to demand a system of standardisation by which particular exemplars might be judged. Kendrick (1872–1954) joined the Victoria and Albert Museum in 1897, becoming the first Keeper of Department of Textiles in 1909. He was a leading authority on medieval textiles and fibres and wrote a number of influential books. One of these, *English Embroidery* (1905), includes a chapter on embroidery from the Anglo-Saxon period and one on what he termed the 'Norman and Early English Period'.[24] This book follows an established template, for example, in 1888 Alan Summerly Cole (1846–1934), secretary to the South Kensington Museum and Director of Art, published his translation of Ernest Lefébure's *Embroidery and Lace: their manufacture and history from the remotest antiquity to the present day*, which contains the chapter, 'From the Christian Era to the Crusades', and in 1894 Frances and Hugh Marshall published *Old English Embroidery: its technique and Symbolism*, which included two similar chapters, 'Anglo Saxon Period' and 'Anglo-Norman Work'.[25]

Many of Kendrick's observations are still pertinent, although his perspective betrays a 19th-century sensibility. He highlights that the production of a particular piece could take a number of years and, using documentary sources, demonstrates that surviving pieces of embroidery are outnumbered by those that have been lost. His work is one of the earliest to analyse early medieval embroidery in its own right, and embed it in the overall history of English medieval embroidery. Christie's work is clearly influenced by Kendrick's methodology, incorporating his ideas but developing her own use of documentary sources to form an interdisciplinary approach to the study of embroidery. The work of the French curator of medieval embroidery, Louis de Farcy, can also be seen as an important influence on Christie, whose method of setting out her work bears the hallmark of his presentational style.[26] Finally, from the embroidery practice-oriented strand of writers, May Morris, who

---

23 L. Cluckie, *The Rise and Fall of Art Needlework* (Bury St Edmunds: Arena Books, 2008), 106–110; B.M. King, *The Wardle Family and its Circle* (Woodbridge: Boydell Press, 2019), 40–83; L.Hulse, 'Elizabeth Burden and The Royal School of Needlework', *The Journal of William Morris Studies*, 22 (2014), 22–34 (22, 23).
24 A.F. Kendrick, *English Embroidery* (London: George Newnes, 1905).
25 A.S. Cole, ed. and trans., E. Lefébure, *Embroidery and Lace* (London: H. Grevel and Co., 1888); F. Marshall and H.Marshall, *Old English Embroidery* (London: Horace Cox, 1894).
26 M.L. de Farcy, *La Broderie de XIème siècle jusqu'à nos jours d'après des spécimens authentiques et les anciens inventaires*, 3 vols (Angers: Belhomme, Libraire-Éditeur, 1890–1919).

studied medieval embroideries and writings on medieval embroidery, can be seen as an influence on Christie through her extensive visits to study actual pieces and her technical knowledge.[27] May Morris's book, *Decorative Needlework* (1893), which focuses on describing embroidery stitches used by contemporary embroiderers, is also of note because the style of minimal text with hand-drawn illustrations used as a form of teaching aid, has become the standard for nearly all such publications subsequent to it.[28]

Also Arts and Crafts influenced, the work of Christie herself (1872–1953) can be seen as drawing together and substantially advancing embroidery research. Her husband, Archibald Christie, who trained as an artist and designer, had become an architect by the time they were married and had collaborated on at least one occasion with Lethaby. The couple also knew Kendrick. Grace Christie's ground-breaking publication is *English Medieval Embroidery* (1938) but the results of microscopic and technical analyses on the stole and maniple from the tomb of St Cuthbert, written up prior to this, are also important. In the course of outlining the results that prove the materials to be silk, she prioritises terminology that is used by textile archaeologists today to describe the threads: 'the silk threads ... have no twist, are used in different thicknesses, and for outlining purposes are coarser than for fillings,' and the gold is solid and formed from 'a strip from 1/90 to 1/150 inch in width and about 1/7000 inch in thickness'.[29]

The focus of *English Medieval Embroidery* is *Opus Anglicanum*, but Christie also included a detailed analysis of the Cuthbert embroideries. The principle of selection she uses is ecclesiastical, and on this basis the Bayeux Tapestry is excluded. The book lacks a bibliography, sources are sometimes patchy and judgements occasionally miss the mark.[30] On the whole, however, the author's strengths in technical observation and analysis of the embroideries give the work a rigour that means it stands out from its predecessors. Christie was before her time in her methodology and analysis. She considered embroidery as an art and a technical skill. She was the first to inform the reader about the wider contexts in which English medieval embroideries were produced. She methodically discussed technical aspects of embroidery production including design, materials, stitch conventions and working practices. She also placed *Opus Anglicanum* in a wider material context, dealing, for instance, with how it was viewed abroad. Finally, she treated the embroiderers

---

27 L. Hulse, '"When Needlework was at its very finest": *Opus Anglicanum* and its influence on May Morris', in *May Morris: Art and Life. New Perspectives*, ed. L. Hulse (London: Friends of William Morris Gallery, 2017), 87–110.
28 M. Morris, *Decorative Needlework* (London: Joseph Hughes, 1893).
29 G. Baldwin Brown and [Mrs] A.Christie, 'S. Cuthbert's Stole and Maniple at Durham', *The Burlington Magazine*, 23 (April 1913), 2–7, 9–11, 17 (5).
30 Christie misidentified the ground fabric. She thought the gold threads were stitched to warp threads only. For discussion of this see: E. Plenderleith, 'The Stole and Maniple (a) the technique', in *The Relics of St Cuthbert*, ed. C.F. Battiscombe (Oxford: Oxford University Press, 1956), 375–396 (382).

themselves with due consideration, including an appendix of named embroidery-workers of all ranks, as opposed to just those from elite circles. This distinguished her work from previous authors who had focused on queens and aristocratic women, thereby romanticising the production of embroidery. She provides a bridge between the early antiquarian circle of research and influence, and later, more scientific approaches, being the first researcher to engage with embroidery at a technical level. Professional textile archaeology as a sub-discipline of archaeology was developing at the same time that Christie was active. It created a new method for textile analysis and was pioneered by a mother and daughter, Grace and Elisabeth Crowfoot. By using her wide knowledge of textiles from across Europe, Grace Crowfoot, née Hood (1877–1957) was able to place British archaeological textiles in their wider design, production and material contexts. In so doing she showed that textiles could inform wider issues within archaeological, art historical, and social and material cultural research. Her only work on early medieval embroidery, concerning the fragment of embroidered gold-work from the Basilica Ambrosiana, Milan, was published in 1956, in the major volume on the Cuthbert embroideries edited by C.F. Battiscombe.[31]

After Grace Crowfoot's death, her daughter Elisabeth (1914–2005) went on to dominate the field and raise the profile of archaeological textiles. Just four of her publications analyse extant embroideries from the early medieval period, and embroidery tends to be an adjunct of the textiles under consideration as opposed to her work's main focus.[32] However, her ability to dissect stitch production and her detailed drawings remain an invaluable aid to researchers, although not all her work stands up to close scrutiny. Also during this period, Elizabeth Plenderleith, the wife of Harold Plenderleith (who established the British Museum's conservation programme), wrote the technical report for the Battiscombe volume.[33] The author states that she was involved in examining the Cuthbert embroideries when they were at the British Museum for cleaning, where their dismantling would have enabled detailed study of the materials and techniques of construction.

---

31 G.M. Crowfoot, 'Note on a Fragment of Embroidery from the Basilica Ambrosiana in Milan', in *Relics of St Cuthbert*, ed. Battiscombe, 392–394.
32 The four early medieval embroideries Elisabeth Crowfoot analysed can be found in: E.G. Crowfoot, 'Appendix I. Objects found in Cremation Heath, Mound II', in M. Posnansky, 'The Pagan-Danish Barrow Cemetery at Heath Wood, Ingleby: 1955 excavations', *Journal of the Derbyshire Archaeological and Natural History Society*, 76 (1956) 40–56 (52–53); E. Crowfoot, 'The Textiles', in *The Sutton Hoo Ship-Burial volume 3*, eds R. Bruce-Mitford and A.C. Evans (London: British Museum Publications, 1983), 404–479; E. Crowfoot, 'Personal Possessions: textiles', in *Object and Economy in Medieval Winchester*, ed. M. Biddle, Winchester Studies, 7.ii (Oxford: Clarendon Press, 1990), 467–493; Crowfoot, 'Textile Fragments'. For a list of works Elisabeth Crowfoot produced for textiles from the early medieval period see, E. Coatsworth and G.R. Owen-Crocker, *Medieval Textiles of the British Isles AD 450–1100* (Oxford: Archaeopress, 2007), 66–76.
33 Plenderleith, 'The Stole and Maniple (a) the technique'.

The approaches taken by the Crowfoots and Plenderleith make useful comparators. Whereas Plenderleith works through information in a scientific manner and presents data in tabular form, Grace Crowfoot's work on the Milan fragment is more descriptive and freeform: measurements are included but remain embedded in the text. Plenderleith's work has the advantage of incorporating Crowfoot's attention to details such as weave construction and thread count at the same time as formatting data for easy comparison across several embroideries. However, she does not integrate this with wider historical contexts as Christie and the Crowfoots did, leaving this instead to specialists in art history, though this is likely to be in part a function of the way work was divided between contributors in the Battiscombe volume.

Researchers from the later 20th century into the present have built on these foundations and innovated. For instance, the work of textile archaeologist Penelope Walton Rogers employs up-to-the-minute scientific and technological approaches, detailed descriptive analysis and attention to the wider textile context. A number of curators and conservators have also made important contributions to study of early medieval embroidery. Frances Pritchard (previously of The Whitworth, Manchester) has contributed important research on archaeological textiles, including recent work on textiles from Viking Age Dublin, and collaboration with the independent scholar Hero Granger-Taylor and Louise Mumford at the National Museum of Wales, Cardiff, on the Llangorse textile, a large piece of burnt fabric found folded in Llangorse Lake, which is believed to have been a tunic and possibly incorporated embroidery.[34] Sylvette Lemagnen, Curator of the Bayeux Tapestry between 1989 and 2017, has done much to publicise new research on the Tapestry to both general and academic audiences, including the work of Isabelle Bédat and BéatriceGirault-Kurtzeman, the two conservators who wrote up the results of the Bayeux Tapestry's conservation during winter 1982–1983. This brought to light for the first time the existence of a previously unnoticed eighth seam near the end of the Tapestry as a result of study of the reverse of the textile.[35]

Art historians have been able to place early medieval embroidery in a wider art historical context. C.R. Dodwell's groundbreaking book, *Anglo-Saxon Art: a new perspective* (1982), one of the most important surveys of the sources of early medieval art, is of particular note here.[36] His discussion of embroidery-workers, which lists documented individuals, was used as evidence for general trends within craft producing circles. He suggests that in the eyes of the Anglo-Saxons, those who produced gold-work embroidery were the female equivalent of goldsmiths and would have been held in high esteem.

---

34 For which Mumford is curator and conservator.
35 I. Bédat and B. Girault-Kurtzeman, 'The Technical Study of the Bayeux Embroidery', in *The Bayeux Tapestry: Embroidering the Facts of History*, eds P. Bouet, B. Levy and F. Neveux (Caen: Presses Universitaires de Caen, 2004), 83–110.
36 C.R. Dodwell, *Anglo-Saxon Art: a new perspective* (Manchester: Manchester University Press, 1982).

Academics working in Anglo-Saxon studies, an interdisciplinary field with a focus on early medieval material culture, have also played a part in textile and embroidery research developments, for instance, in the interdisciplinary collaborative research involving the Maaseik embroideries which took place between 1979 and the mid-1980s. The team of British and Belgian specialists included: Penelope Walton Rogers, John Peter Wild and Donald King, all authorities on textiles; Hero Granger-Taylor, a specialist on silk; David Wilson, Leslie Webster and James Graham-Campbell, experts in Anglo-Saxon art; and the authors of the two articles associated with the project, Mildred Budny and Dominic Tweddle, researchers in Anglo-Saxon manuscripts and Anglo-Saxon archaeology respectively.[37] Although the project's scientific approach to examining and measuring materials (thread counts per square cm, the twist and ply of embroidery thread and the width of gold foil wrapped around a silk core to form gold thread), is indebted to the work of Walton Rogers, it is interesting to note that when technical embroidery terms and techniques are discussed, the authors refer back to Christie. Various stitches are highlighted, but only the technique used to form couched-work is explained.

Gale R. Owen-Crocker has published numerous works on the Bayeux Tapestry, several of which also highlight aspects of the embroidery's construction or stitch work. But it was not until 2005 that the article by Elizabeth Coatsworth, 'Stitches in Time', brought together information about all then known English-made early medieval embroideries, and placed them into an embroidery context with pieces from Continental Europe. This drew scholars' attention to the corpus of early medieval embroidery for the first time and confirmed its status as part of early medieval material culture.[38]

Coatsworth also published a thought provoking paper on the Cuthbert embroideries, using them as a material focus to help explain broader contexts of King Edward the Elder's reign (d. 924), and provide an insight into his household.[39] Her research enabled her to deduce that the embroidery known as Maniple II had originally been two pieces of textile that had at some point been joined together. She concluded that Maniple II was originally meant to be seen from both sides,took the logical step of looking for similar textiles within manuscript illuminations of the same date, and found depictions of such textiles worn by Anglo-Saxon women as cloak ribbons and headbands. It is unknown when this textile was stitched together into one piece; however, if Coatsworth is correct, this embroidery was originally a

---

37 M. Budny and D. Tweddle, 'The Maaseik Embroideries', *Anglo-Saxon England*, 13 (1984), 65–96 (66–67) and M.Budny and D. Tweddle, 'The Early Medieval Textiles at Maaseik, Belgium', *The Antiquaries Journal*, 65 (1985), 353–389.

38 Owen-Crocker and Coatsworth have collaborated on a number of useful works on textiles. See their co-authored volume, *Medieval Textiles of the British Isles*, which provides researchers with an important resource on textile production and scholarly work on textiles. Their most recent collaborative work, co-edited with Maria Hayward, *Encyclopedia of Medieval Dress and Textiles of the British Isles c. 450-1450* (Leiden: Brill, 2012), added further to the scholarly corpus, collating information on as many aspects of textiles and associated tools and uses as possible.

secular garment, but became a religiously significant object when it was donated to the tomb of St Cuthbert. Coatsworth's work makes a convincing case that the discovery of an embroidery in a religious setting does not prove it was originally made for such a purpose.

Finally, the work of professional embroiderers can provide valuable insights into early medieval embroidery. Of particular note are the efforts of Jan Messent to analyse the Bayeux Tapestry from an embroiderer's perspective and the work of Helen M. Stevens, who between 1985 and 1986 undertook a project to recreate one of the arcade strips from the Maaseik embroideries.[40] While the resulting publications lack academic frames of reference, the work of each embroiderer demonstrates that practical application and experimentation are useful tools. For example, Stevens was able to hypothesise the time it may have taken to embroider an arcade strip (see p. 120 for discussion). Embroidering samples gives insight into working methods and practical needs that scholars working with texts or the artefacts alone cannot appreciate. The trained embroiderer can see beyond the object in question to its constituent parts and the logistical practicalities involved in producing the embroidery. When combined with the skills of the academic researcher utilising historical and archaeological research methods, a far greater variety of data can be gathered together and deployed.

The shape of the book

In light of this, the relevance of my own skill set to this particular project, which derives from a Royal School of Needlework apprenticeship combined with academic training as an archaeologist, merits mention. It means that this study can range freely between detailed technical observations of individual embroideries and informed contextualisation of the cohort of embroidery under investigation within its early medieval social environment. It enables me to use the information gathered, together with a survey of the sources and archaeological data from the period, to hypothesise the production methods and training undertaken by the embroidery-workers, and the organisational strategies utilised at production sites and workshops. Moreover, a greater level of understanding is possible not only with regard to production, and the development of style and technique, but concerning the wider social and political contexts within which early medieval embroidery played an important material cultural role: creating relationships, enhancing memory, and cementing and strengthening affiliations, power and authority.

---

39 E. Coatsworth, 'The Embroideries from the Tomb of St Cuthbert', in *Edward the Elder 899–924*, eds N.J. Higham and D. Hill (London: Routledge, 2001), 292–306.
40 J. Messent, *The Bayeux Tapestry Embroiderers' Story*, 2nd edn (Tunbridge Wells: Search Press, 2010), and H.M. Stevens, 'Maaseik Reconstructed: a practical investigation and interpretation of 8th-century embroidery techniques', in *Textiles in Northern Archaeology,* eds Walton and Wild, 57–60.

The chapter that follows, called 'The data and the difficulties', presents the statistical data arising from the research project and discusses the conclusions to be derived from assessing the corpus as a whole, together with this approach's limitations. Discussion is then extended to encompass the challenges relating to the archaeological finds context in which many pieces in the corpus were discovered, demonstrating through reference to specific examples, why studying medieval embroidery is notoriously hard. The discussion also introduces readers to aspects of a number of the pieces considered in later chapters. Chapter 3, 'Kempston: the biography of an embroidery', uses detailed consideration of the Kempston fragment to illustrate how in-depth study can aid and enrich understanding, not only of the object itself, but its milieu. Discussion utilises object biography theory (of which I say more below), to demonstrate the way data from a variety of sources can be combined to begin to piece together the life of an object. Situating the object in this way suggests further directions for study: principally its functioning as a significant part of Anglo-Saxon material culture and the evolution of embroidery production over time. Thus Chapter 4, 'Embroidery and Anglo-Saxon society', broadens out the discussion in order to build on detailed findings specific to individual pieces and situate embroidery more clearly in the context of the material culture and social world of the Anglo-Saxons, exploring its functions and meanings. Chapter 5, 'Early medieval embroidery production in the British Isles', uses the diverse material of preceding chapters to reach an understanding of the stages in embroidery production's development in the British Isles, and finally, the concluding chapter, 'Embroidery in context', brings together the book's findings to assess what we have learnt from this study.

Scope of the project

The pieces in this study range from small, mineral 'replaced' embroideries (pieces that no longer consist of their original materials), such as an extremely fragmented early 7th-century cuff from Sutton Hoo (known as 'Sutton Hoo B') (Fig. 5) to the very large, of which the 68.38 m embroidered hanging known as the Bayeux Tapestry is the best known. While commonalities of design, materials and/or stitch work may be shared across the corpus, each individual piece is unique, meaning that study of the corpus must pay due attention to the individuality of each embroidery, achieved here using an interdisciplinary approach.

To provide a clear frame of reference for the exemplars, embroidery is here restricted to its most limited formal expression, that is, decorative work created with sewing thread, or decoratively attaching other threads such as gold, to a pre-woven ground fabric. Pre-formed items that are attached to the ground fabric with functional stitches (for example, smaller pieces of fabric, bands and plaits, which are known as appliqué), are not included in the corpus surveyed, unless they are associated with the embroidery under discussion. Decorative woven techniques such as brocading and soumak, which require the thread to be interlaced into the fabric as it is being created, are excluded by a similar rationale, as are tablet-woven bands,

1. Introducing Anglo-Saxon embroidery   17

Figure 5. Sutton Hoo B: Line drawing of the embroidery (50 × 30 mm), © Trustees of the British Museum

which are created by twisting threads together and securing them by a warp thread to form a patterned band that can be attached to soft furnishings or clothing.[41] Applied metalwork – pulled metal wire woven into flat or 3D shapes such as brooches and miniature bells or balls, before being sewn on to fabrics – is also outside the scope of the survey. The Catalogue contains a brief description and the vital statistics for each piece of embroidery in the corpus. Items in the catalogue are listed according to their find location. A full description of each type of embroidery stitch along with basic textile terminology can be found in the Glossary. The Table of Surviving Insular Embroideries (Table 1A–G), which provided an important organising structure for thinking about the 43 embroidered items in the main corpus, is also included.

Once all published and archival data for individual embroideries had been collated in catalogue and tabular form, as many visits as possible were conducted in order to view the pieces in situ. Seventeen of the 44 pieces were available for viewing.[42] Of the

---

41 For further details on these techniques see Emery, *Primary Structures of Fabrics*, 68–69, 215–216, 225, 251.
42 Note the discrepancy between 43 embroidered items and 44 pieces. This is because the small piece cut from the Cuthbert embroideries stole is not given a separate catalogue entry since it is essentially the same embroidery as the stole, however, physically it counted as a separate piece to visit (especially owing to the fact that the stole itself was unavailable in storage).

Figure 6. Alfriston: Two buckles with embroidery attached (measurements not available), from Griffith and Salzman (1914), pl. xv, 16

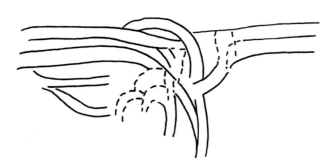

Figure 7. Line drawing of the Worthy Park embroidery (embroidered layer: c. 25 × 15 mm), © Alexandra Lester-Makin, after Srahan (2004)

twelve embroideries located across the United Kingdom (and viewable), four are in London: the embroidered seam and cuff from Sutton Hoo and the fragment from Kempston in the British Museum,[43] and a partial shoe with an embroidered top band, held in the Museum of London archive.[44] The others are the possible tunic with embroidery found at Llangorse, in the National Museum of Wales, Cardiff;[45] a nearly complete shoe in the Herbert Art Gallery and Museum, Coventry (Pl. 6b);[46] a fragment of metal embroidery held by the Derby Museum and Art Gallery (Pl. 7a, b);[47] fragments, catalogued here as a single piece, which may originally have been a stole and maniple, stored at Worcester Cathedral Library;[48] and, at the Durham University Conservation Centre a small piece cut from the Cuthbert embroideries' stole (Pl. 7c),[49] a band from the Cuthbert embroideries (Pl. 8a),[50] and two fragments from the tomb of Bishop William of Calais (Pl. 8b, 9a).[51]

The other five items made available for study are either stored, or on display, outside the British Isles and Ireland. These are the Bayeux Tapestry, which is housed at the Musée de la Tapisserie de Bayeux, France; a fragment of embroidery held at the Basilica Ambrosiana in Milan, Italy (Pl. 9b); and the Maaseik embroideries – a group of eight embroidered pieces, which I have here catalogued as three sets because of

---

43 Catalogue entries Sutton Hoo A and B, and Kempston.
44 Catalogue: London A.
45 Catalogue: Llangorse.
46 Catalogue: Coventry.
47 Catalogue: Ingleby.
48 Catalogue: Worcester.
49 Catalogue: Durham D. The stole fragment is listed in the Catalogue as part of Durham D, the 'stole with embroidery', which is housed at Durham Cathedral, because it is known and has been technically verified to be a fragment of the main garment.
50 Catalogue: Durham A.
51 Catalogue: Durham F and G.

1. Introducing Anglo-Saxon embroidery

Figure 8. Mitchell's Hill wrist clasps (AN1909.487.[i]). Wrist clasp with some traces of textile (tablet-woven band: 28 × 13 mm), Image © Ashmolean Museum, University of Oxford

the likely common origins of pieces within each of my three subgroups. These pieces from the embroidered textile of the supposed *casula* of St Harlindis are displayed at the Church of St Catherine, Maaseik, Belgium (Pl. 10).[52]

Visits were not possible for the other 27 embroideries, either owing to problems locating the items or restrictions on access. The embroideries from Alfriston (Fig. 6), Worthy Park (Fig. 7), Mitchell's Hill (Fig. 8) and Utrecht are missing. While every attempt has been made to locate them, they have not yet been found, suggesting they may have been discarded. In addition, York Archaeological Trust and Yorkshire Museum were unable to locate the embroidered pouch and sleeve edge from York. This was also the case with regard to a fragmentary shoe which is the responsibility of the Museum Store in Winchester (Fig. 9).[53]

---

52 Catalogue: Bayeux, Milan and Maaseik A, B and C. There are eight separate embroideries at Maaseik; however, they are grouped as three sets because there is evidence that the two strips with roundel designs were probably an autonomous set, and the same is true of two strips with arcade designs, while four monograms were most likely attached to a single object after they were first made.

53 Catalogue: Alfriston, Worthy Park, Mitchell's Hill, York A, York B and Winchester. Alfriston is a possible impression of embroidery in rust on the inner side of the tabs of an iron buckle and was recorded in A.F. Griffith and L.F. Salzmann, 'An Anglo-Saxon Cemetery at Alfriston, Sussex', *Sussex Archaeological Collections, Relating to the History and Antiquities of the County*, 56 (1914), 16–51; Worthy Park is a replaced embroidery fragment and is recorded by E. Crowfoot, 'Chapter 5: The Textile Remains', in *The Anglo-Saxon Cemetery at Worthy Park, Kingsworthy Near Winchester, Hampshire*, Oxford School of Archaeology Monograph 59 (Oxford: Oxford University School of Archaeology, 2003), 192–195; Mitchell's Hill is a possible embroidered tablet-woven band, recorded in G.Crowfoot, 'Anglo-Saxon Tablet Weaving', *The Antiquaries Journal*, 32 (1952), 189–191.

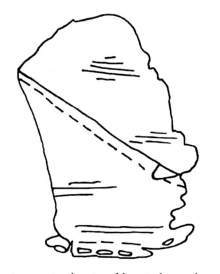

Figure 9. Line drawing of the Winchester shoe, © Alexandra Lester-Makin, after Thornton (1990)

Four pieces from Durham: the second band, a possible girdle or ribbons, the stole, and the maniple, were inaccessible due to renovations taking place at the Cathedral and then because of their vulnerable state.[54] Twelve silk and wool fragments from the Oseberg Ship are stored at the Ship Museum in Oslo, Norway, but were also unavailable due to building work (Figs 10–17, Pl. 11).[55] Restricted access also prevented viewings of a decorated fragment of embroidery at the National Museum of Dublin, and the Orkney hood at the National Museum of Scotland in Edinburgh[56] and limited me to viewing just one of the three fragmentary shoes stored at the Museum of London archive.[57] It is hoped that the recovery of lost items and changes in viewing policies may allow further study to supplement research for a future edition. In the meantime the study makes use of existing published literature, archaeological reports and archival sources relating to these items.

Although the Catalogue and Tables 1A–G bring together the 43 embroideries and fragments of embroidery surveyed in this book, it is pertinent here to acknowledge further categories of pieces that have not been included in the survey. These pieces inhabit a middle-ground between the categories excluded by my definition discussed above (see pp. 16–17) and the 43 extant pieces. They are tabled separately in the two tables in Appendix 3, and occasionally provide useful contextual material for discussion. First, Table 2: Loose gold thread lists surviving thread likely to have been used for embroidery, but for which there is no firm evidence. We can hypothesise that surviving loose gold threads were once woven into a band or used for embroidery by studying the type and positioning of bends and dents left in the filament by the weaving and sewing processes.[58] This form of archaeological analysis has been performed on all the examples included except for Southampton (Southampton Ga), a reel of gold thread found in a cess pit (Fig. 18).[59] This is included in the same

---

54 Catalogue: DurhamB, C, D and E.
55 Catalogue: Oseberg A–L.
56 Catalogue: Dublin and Orkney.
57 Catalogue: London B and C.
58 Pers. comm. Frances Pritchard, formerly of The Whitworth (2015).
59 P. Walton Rogers, 'Gold Thread', in *The Origins of Mid-Saxon Southampton: Excavations at the Friends Provident St Mary's Stadium 1998-2000*, eds V. Birbeck, R.J.C. Smith, P. Andrews and N. Stoodley (Salisbury: Wessex Archaeology, 2005), 68–69.

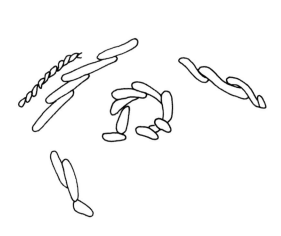

Figure 10. Line drawing of Oseberg A (c. 30 × 30 mm), © Alexandra Lester-Makin, after Stranger (1979)

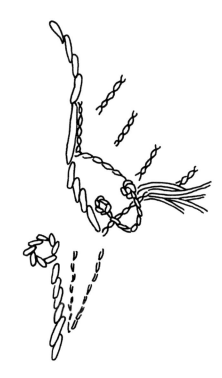

Figure 11. Line drawing of Oseberg B (measurements not available), © Alexandra Lester-Makin, after Stranger (1979)

table because it represents the metal thread type used in embroideries. It seems reasonable to assume that what we have been left with in this case is a reel that may have been earmarked for embroidery, weaving or brocading, but had not yet been used, while the loose threads are examples of threads that had been used one or more times and then (probably) set aside for recycling into new products.

Also included in Appendix 3 is Table 3: Pieces with no surviving thread – textiles and leather with possible embroidery holes but no thread evidence. The holes are thought to be evidence of embroidery from their positioning in the ground fabric and relative to each other, and associated fabric distortions, which provide the specialist observer with clues to where thread used to lie, the direction in which it was pulled, and the overall pattern of use. Archaeological textile experts including Grace and Elisabeth Crowfoot have used this information to suggest stitch reconstructions, from which they can hypothesise decorative or functional use. The information for both the gold threads and textiles and leather with holes was taken from published reports or utilised personal communication with specialists analysing the objects. Although this data does not form part of the main survey, findings have been included as supporting evidence where applicable. Trips to see some of the items included in Appendix 3 were integrated with my research where they fitted alongside other visits: the three textile fragments with no surviving threads in London (one from Buckland and two from Sutton Hoo); two each at the York Archaeology Trust and Yorkshire

Figure 12. Oseberg C (measurements not available), © Museum of Cultural History, University of Oslo, Norway

Figure 13. Oseberg E (56 × 30 mm), © Museum of Cultural History, University of Oslo, Norway

1. Introducing Anglo-Saxon embroidery

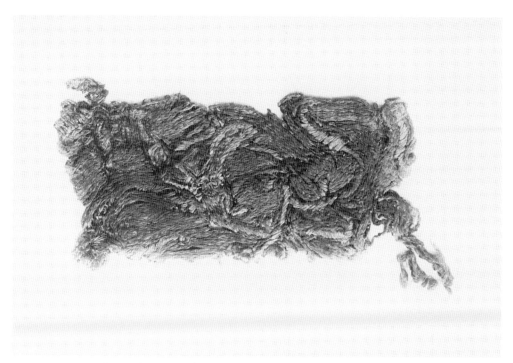

Figure 14. Oseberg F (65 × 30 mm), © Museum of Cultural History, University of Oslo, Norway

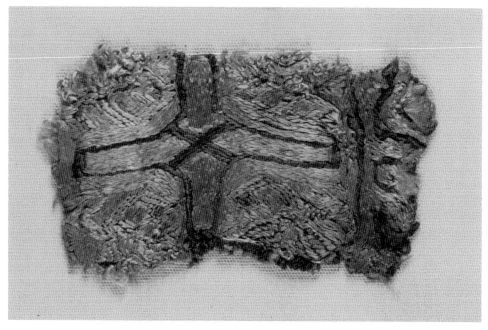

Figure 15. Oseberg G (33 × 54 mm), © Museum of Cultural History, University of Oslo, Norway

Figure 16. Oseberg H (45 × 40 mm), © Museum of Cultural History, University of Oslo, Norway

Figure 17. Oseberg I (38 × 25 mm), © Museum of Cultural History, University of Oslo, Norway

Museum (three of which are fragmentary shoe parts); and the loose fragments of gold embroidery from Repton, Derbyshire, and Winchester, held in safe-keeping by Professor Martin Biddle in Oxford.

On each research trip the embroideries were recorded photographically using a Nikon D3200 digital camera, microscopically with a Veho Discovery VMS-004 USB microscope, and in written form. Fabric, threads and other surviving materials were analysed by eye and recorded in note form. Measurements were taken for the overall size, the elements of each design and individual stitches. Connections and similarities were noted as were any potential irregularities. Pertinent archival material and available associated artefacts were also examined.

Using object biographies

It became clear during research visits that the standard of records varied widely: the level of detail, the attention to technical and design elements, and whether or not wider embroidery contexts were provided. It became of paramount importance that the history of each piece was reconstructed as far as possible so that it could be meaningfully considered alongside other examples. Embroidery and its early medieval social context cannot be compartmentalised: the inextricable links between, for instance, owner and embroidery; producer, designer, worker and patron; giver and receiver; and social and material conventions, mean that embroidery and its context each shed light on our understanding of the other. Strategies associated with object biography theory were utilised to achieve this.

Figure 18. Reel of gold thread from Southampton, image: Elaine Wakefield, © Wessex Archaeology

Over the last twenty to thirty years, the study of material culture from within the discipline of archaeology has drawn on approaches developed in other fields, including anthropology, the social sciences and cultural theory, to inform understanding of particular societies, such as that of early medieval England.[60] The study of prehistory has proven especially fertile ground for theoretical approaches because there are no documentary records to help scholars interpret the archaeological evidence: researchers are reliant on sophisticated analysis of archaeological artefacts, landscape and excavations to help them interpret prehistoric societies. These theories can also be utilised in the study of those periods where the survival of documentary sources is sporadic rather than non-existent, including the early medieval period.

Object biography theory has proved to be a particularly fruitful approach in enabling greater understanding of early medieval material culture and society. For instance, in a study carried out at the 7th-century cemetery at Sutton Hoo, L. Alcock, Angela Evans, J.D. Richards and C.J. Scull used the object evidence, particularly the

---

60 For an introduction to developments, see A. Jones, *Archaeological Theory and Scientific Practice* (Cambridge: Cambridge University Press, 2002); C. Knappett, *Thinking through Material Culture: an interdisciplinary perspective* (Philadelphia: University of Pennsylvania Press, 2005); J.H. Arnold, *What is Medieval History?* (Cambridge: Polity Press, 2008), 57–85.

material found within Mound 1 (a ship burial), to argue that people from across all strata of early medieval society used objects in sophisticated ways, often attributing both functional and symbolic roles, sometimes simultaneously. This, the authors argued, facilitated the development of particular expressions of cultural identity.[61]

Object biography's use as a methodological approach was first proposed by Igor Kopytoff, an anthropologist interested in social structure and change, in the mid-1980s.[62] Kopytoff proposed that it was possible to trace transformations in the status of a thing – to a commodity embedded in patterns of human exchange, and, in some cases, back to an inert object – by writing its biography, and by doing so, use it as an expression of the culture of which it was a part. By asking similar questions of the artefact as one would when writing the biography of a person, the scholar gains an understanding of the processes and events involved in its 'life'. Kopytoff went on to suggest that by asking such questions researchers could begin to unlock more subtle meanings about the culture within which things are located, such as, for instance, how a particular culture defined, or redefined, an object adopted from another culture.[63] The focus of this theory is the object and what it can tell researchers about itself and the culture from which it came. As Andrew Jones noted, prior to Kopytoff's work the notion of biography for objects only came about when the item became a gift, and thus only applied in gift-based economies.[64] Kopytoff's more nuanced approach redefined and extended the idea of biography, treating the object as an entity in its own right and relating it to the culture from which it came.

The anthropologist Alfred Gell took Kopytoff's ideas further, proposing that we understand an object as endowed with a life, or agency, of its own, arguing that humans have very little input into this agency once the object has been produced.[65] From the late 1990s, archaeologists also began to see object biography's potential, particularly in the field of prehistory. By the time the journal *World Archaeology* dedicated a volume to the subject in 1999 the theory had also been adopted by medieval scholars. Of particular note for the study of embroidery are instances of early medievalists who

---

61 L. Alcock, 'Message from the Dark Side of the Moon: western and northern Britain in the age of Sutton Hoo', in *The Age of Sutton Hoo: the seventh century in north-west Europe*, ed. M.O.H. Carver (Woodbridge: Boydell, 1992), 205–215; and, in the same volume, J.D. Richards, 'Anglo-Saxon Symbolism', 131–147; and C.J. Scull, 'Before Sutton Hoo: structures of power and society in early East Anglia', 3–22. See also A. Evans, 'Seventh-century Assemblages', in *Sutton Hoo: an Anglo-Saxon princely burial ground and its context*, ed. M. Carver, Reports of the Research Committees of the Society of Antiquaries of London, 69 (London: British Museum Press, 2005), 201–282.
62 I. Kopytoff, 'The Cultural Biography of Things: commoditization as process', in *The Social Life of Things: commodities in cultural perspective*, ed. A. Appadurai (Cambridge: Cambridge University Press, 1986, repr. 2011), 64–91.
63 Kopytoff, 'Cultural Biography', 66–67.
64 Jones, *Archaeological Theory*, 83.
65 A. Gell, *The Art of Anthropology: essays and diagrams*, ed. E. Hirsch (London: Athlone Press, 1999). Gell died before he was able to refine his work, which has caused a great deal of debate since publication.

extend the idea of the object life cycle from the original 'use and discard' model to the 'multiple life cycle' model, extending from re-use to post-excavation. For instance, John Moreland demonstrates that the Bradbourne cross has been 'reincarnated' throughout history according to what people thought it represented at particular periods. He argues that the cross's reincarnations, uses, abuses and 'preservation' are all visible to be read when one knows what to look for.[66] The cross in its present form tells not only its own biography but also those of the generations of people who have been in contact with it. Similarly, Cornelius Holtorf, discussing the biography of an excavated pot sherd, suggests that biography does not stop with deposition; instead the sherd acquires a 'second life' when the object is rediscovered.[67]

This is highly significant with regard to a number of the early medieval embroideries. The fragments researchers see today are not what people saw – either literally or metaphorically – at the time they were made, used, recycled or deposited. Take, for instance, the embroideries placed in Saint Cuthbert's tomb in c. 934. Even during the course of the relatively recent period of time since their rediscovery in the 19th century, these embroideries have undergone a number of conservation procedures which have radically altered how they look. Likewise, 20th-century viewers have not interpreted the embroideries in the same way as they did on discovery in 1827, as Coatsworth's persuasive reinterpretation of Maniple II as cloak ribbons underlines, and the same may be true for 21st-century viewers now the embroideries have been redisplayed.

Ten years after the first volume of *World Archaeology* appeared, Jody Joy argued for an extension of the application of object biography theory.[68] He marks part of a broader move towards greater emphasis on context – on using objects to gather data on production techniques and evidence of long-term human interaction.[69] Joy suggested that by adopting a relational approach to interpreting the biography of an Iron-Age mirror, greater insight can be gained. He proposed that the making of the mirror provided a stage for developing and/or cementing relationships, while in return, the mirror itself acted as a marker of social influence for its makers. The mirror's life was not static, nor was it constant; rather it changed as social relationships were set out around it and as a result of it.[70] In this way object and human social relationships are demonstrably entwined, and Joy was suggesting that it was the human group, not just its owner, that was entwined with the object, whether in a closely knit group or as part of a larger social network.

---

66 J. Moreland, 'The World(s) of the Cross', *World Archaeology*, 31 (1999), 194–213.
67 C. Holtorf, 'Notes on the Life of a Pot Sherd', *Journal of Material Culture*, 7 (2002), 49–71.
68 J. Joy, 'Reinvigorating Object Biography: reproducing the drama of object lives', *World Archaeology*, 41 (2009), 540–556 (542).
69 Joy, 'Reinvigorating Object Biography', 542.
70 Joy, 'Reinvigorating Object Biography', 552.

Interpreting embroidery in the same way, an initial set of relationships can be seen as developing within the context in which the raw materials were sourced: a network of producers and merchants linked to the person who commissioned the embroidery. Once the materials had been gathered, a second set of relationships were established linking the commissioner, designer and worker(s). The third set, linking commissioner and recipient, were created once the embroidery was complete. For each set of relationships the context changes: raw materials – variously of animal, vegetable or mineral – originating from different places, possibly different continents, are brought together under one roof. The workshop combines the raw materials together to form a finished embroidery which is passed on to the person who commissioned it. The patron gifts the embroidery to the receiver who may reside elsewhere. It can be argued that the human relationships associated with the embroidery morph to fit circumstance. In turn the status and meanings that the embroidery acquires are fluid, altering according to the context and narrative of the embroidery over the course of its 'life'.[71]

By utilising object biography as a methodological approach, it became possible to tell more detailed life stories for the embroideries in the study. Although complete biographies cannot be written for any of the surviving examples, it is nevertheless possible to analyse them and draw out information that gives a more nuanced understanding to their 'lives' and the close knit groups of people who came into contact with them. In some cases (for example, the Cuthbert embroideries), 'life' after re-discovery by antiquarians offers us fascinating insights into those who subsequently came into contact with the embroideries as well.[72] Such analysis not only reveals the thought processes of later generations, but also how they understood the embroideries and what they perceived them to represent; it places embroideries in the wider social arena and demonstrates their importance across generations.

Despite this potential, there are limitations to the use of object biographies in the context of this study. Since it was not possible to view all of the embroideries, the data for those pieces that were inaccessible is filtered through the words and views of previous researchers: thus not all pieces are equally suitable as case studies. While each embroidery has an entry in the catalogue and Table 1, the examples selected for more detailed biographical discussions are those for which the largest quantity of data has been collected: the viewable pieces offer the most viable – in the sense of the fullest – object biographies. These case studies have been used to structure

---

71 Jones, *Archaeological Theory*, 102.
72 For example, J. Raine, *Remarks on the 'Saint Cuthbert' of the Rev. James Raine, M.A.* (Newcastle: Preston & Heaton, 1828) demonstrates Anglican attitudes towards Roman Catholics when he writes that the tomb of St Cuthbert was opened on 17 May 1827 by the Rev. William Nicholas Darnell, prebendary of the sixth stall of Durham Cathedral, because he wanted to 'open the eyes of the blind deluded papists to the impostures of their church, by demonstrating to them that the body of the patron saint, in defiance of its boasted incorruptibility, is crumbled into dust', (5–6).

more general discussions concerned with trends within embroidery style, design, production and use, which take in those pieces for which less data is available.

Object biography theory itself also has limitations, particularly when questions are posed about the interactions of objects and people within larger social contexts. While Jody Joy extended the idea of the object biography to incorporate social relationships over longer periods of time, all the social interactions he investigated were linked directly in some way to the object. My study's interest in the way embroidery interacts with the surrounding environment, for instance, the deterioration or chemical change that takes place when an embroidery is deposited in the ground is harder to take account of within the framework of a traditionally conceived object biography, which argues the object is 'dead' when buried.[73]

While not overlooking these potential difficulties, my use of object biography here has a broader purpose. It is a way of opening out discussion of early medieval embroidery, especially for those new to its study, and offering a methodology for thinking about textiles that allows us to wrestle productively with a small, somewhat obscure dataset. The first two chapters throw these issues into relief: Chapter 2 focuses on the early medieval embroidery dataset that forms the source material of this study, and the difficulties associated with studying embroidery; in contrast, Chapter 3 takes one particular embroidery, the Kempston fragment, and assembles an object biography as a means of bringing the reader closer to the material culture that is embroidery.

---

73 Tim Ingold's work on meshwork theory is one of a number of theoretical approaches developed in recent years in response to the overly object- and/or people-focused nature of research in social sciences and archaeology, offering a framework that incorporates non-human and non-object aspects of the environment, on an equal footing. See T. Ingold, *Being Alive: Essays on Movement, Knowledge and Description* (London: Routledge, 2011), 91.

Chapter 2

The data and the difficulties

Introduction

Interest in embroidery from early medieval England has been patchy. From the 19th century onwards researchers have been dazzled by examples of the intricate (the Cuthbert embroideries), and the monumental (the Bayeux Tapestry), the less showy examples, such as the small piece of woollen embroidery that forms the Kempston fragment, are little studied. Surviving examples from Wales, Scotland and Ireland are few in number, and have received varying degrees of attention. Archaeological embroidery finds have been analysed by specialists writing reports on the textiles onto which the embroidery was sewn. Some, such as the analysis undertaken on the York sleeve edge (York B), are extremely thorough and detailed but this is not always the case. As a result, the scholarship has remained fragmentary and incomplete.

This chapter works towards changing this, breaking new ground by bringing together and providing introductory analysis for the data gathered over the course of this project, including visual and microscopic analysis, as well as published and archival information. Technical data, including materials, stitches and design elements, is collated in the Table of surviving insular embroideries (see Table 1A–G), supplemented by pie charts, where appropriate, to facilitate easy appraisal. See also the Catalogue (Appendix 1), where key information for each surviving piece is included. The chapter begins, however, by highlighting key difficulties involved in studying early medieval embroidery by providing a brief survey of materials and find conditions and how these interact.

Finds context

The environmental conditions in which an embroidery is recovered make an enormous difference: they dictate to what extent, if at all, the fibres that make up the threads of the embroidery survive. Of the 43 embroideries covered in this study, the majority were discovered in archaeological contexts. Three were discovered during antiquarian excavations (Alfriston, Kempston and Mitchell's Hill) while 25 were found during modern archaeological excavations (Coventry, Dublin, Ingleby, Llangorse, three from London, 12 from Oseberg, two from Sutton Hoo, Winchester, Worthy Park and two from York). The remaining 14 were found in non-excavated settings, the majority of which were shrines or burials in churches: seven from Durham, three from Maaseik, one from Milan, one from Utrecht, one from Worcester and one from Bayeux. The final piece, the Orkney hood, was found by chance in a bog in the parish of St Andrew on Orkney, so it occupies a halfway house – not from an archaeological excavation but nevertheless from an outdoor context that might conceivably have been the site of excavation (see Table 1A: Discovery context). This section is designed to give the reader an appreciation of why so few embroideries from the early medieval period have survived at all. Existing guides to textile excavation focus on excavating and conserving textiles and leather rather than dealing with the specific issues that relate to embroidery, so discussion focuses on the variety of conditions in which early medieval embroideries are found, how this affects the way they survive, and the options for their subsequent treatment.[1]

Material structure

As Table 1A shows, the cohort of embroideries comes from a wide range of dry, damp and burnt environments. The likelihood of survival and the condition of each piece is dependent not only on the fibres from which it was constructed, but how those fibres interact with both the surrounding environment and with other grave goods, where they are present. The results of experiments undertaken by Bill Cooke showed that cotton and woollen textiles become so tender during burial that after only three weeks in a biologically active soil at 20ºC, they can disintegrate under their own weight.[2]

Early medieval embroideries were constructed of cellulose fibres, protein fibres and metals, in the form of gold, silver or gilt thread. Cellulose fibres come from plants

---

1 Key guides to textile conservation are J.P. Wild, *Textiles in Archaeology* (Princes Risborough: Shire Publications, 2003) and S. Landi, *The Textile Conservator's Manual*, 2nd edn (London: Routledge, 1998). For guides dedicated to the excavation and conservation of archaeological textiles see C. Gillis and M.-L.B. Nosch, eds, *First Aid for the Excavation of Archaeological Textiles* (Oxford: Oxbow Books, 2007) and J.M. Cronyn, *The Elements of Archaeological Conservation* (London: Routledge, 1990; repr. 2001). For the conservation of textiles in museum contexts see J.M. Glover, 'Conservation and Storage: textiles', in *Manual of Curatorship: a guide to museum practice*, 2nd edn, ed. J.M.A. Thompson (Oxford: Butterworth-Heinemann, 1992; repr, 1994), 302–339; and, in the same volume, C.V. Horie, 'Conservation and Storage: leather objects', 340–345; and E. Pye, 'Conservation and Storage: archaeological material', 392–426.
2 B. Cooke, 'Fibre Damage in Archaeological Textiles', in *Archaeological Textiles: occasional papers 10*, eds S.A. O'Connor and M.M. Brooks (London: The United Kingdom Institute for Conservation, 1990), 5–14 (5, 9).

and those used in early medieval embroideries are called bast fibres. These are not as pure as cotton fibres, which are made from single hair cells. Instead bast fibres are constructed from lignified (woody) dead cells that grow up the side of plant stems, providing support. The most common example in early medieval embroidery is flax, which is used to make linen fabrics.[3] Protein fibres – of animal origin – differ from cellulose fibres because they are made from strands of keratin and make up the outer layers of skin; wool is such an example. These fibres differ in length depending on the make-up of the keratin fibril cells (thread-like filaments). Wool fibres are fine and short, so that when they are spun, the flattened scales making up the outer layer hook together, binding the short fibres to form longer lengths that become thread. Silk is simpler in construction. It is produced as one long filament that the silk worm uses to construct its cocoon. If this is unwound before the moth eats its way out, it forms a single long length of thread which may be spun if desired; if the larva develops into a moth and bites through the cocoon, the length of filament is reduced and the shortened fibres must be spun together to create a longer thread.[4] Leather is made from the corium (thick outer layer) of an animal skin, which consists of the fibrous protein, collagen. Leather varies in thickness and quality depending on the animal species and its age when slaughtered. Processing shortens the collagen fibres, altering the texture and physical qualities of the finished leather, while tanning waterproofs it through a chemical reaction between the tanning agent and collagen.

Metal threads are constructed very differently to those of plant or animal origin. They provide a distinct group of embroidery threads which survive in three forms from the early medieval period. The first of these is a combination of gold metal wound round a core fibre. The gold is first hammered into thin sheets and cut into strips. Surviving examples of core fibres from this period include silk and horse hair (both being types of protein fibre). The second form is a silver-gilt thread produced when a thin strip of silver is coated with a layer of gold and the whole wrapped round a core of either silk or animal hair. The third type consists of a rod of solid silver that was pulled through a drawing plate to form a thread.[5]

Survival of materials

Embroidery becomes subject to processes of deterioration long before it becomes part of the archaeological record or is placed in storage. Indeed, damage may occur to plant fibres as they grow – perhaps owing to drought conditions; or to animal fibres if the creature concerned becomes ill or has a poor diet. The fleeces of early species of sheep would naturally weaken and break before shedding, and were susceptible to ultra-violet

---

3 Cronyn, *Elements of Archaeological Conservation*, 284; Landi, *Textile Conservator's Manual*, 22–24.
4 Cronyn, *Elements of Archaeological Conservation*, 284–285; Landi, *Textile Conservator's Manual*, 8–9; M. Vedeler, *Silk for the Vikings*, Ancient Textiles Series, 15 (Oxford: Oxbow, 2014), 48–49, 52.
5 For further information see E. Coatsworth and M. Pinder, *The Art of the Anglo-Saxon Goldsmith* (Woodbridge: Boydell Press, 2002), 64–101 (91–94); Landi, *Textile Conservator's Manual*, 12.

damage. In fact, light – ultra-violet in particular – is the cause of a continuous process of molecular breakdown for all cellulose and protein based fibres during their life span.

Processing of fibres also results in damage of sorts. Cellulose fibres may become weak and liable to split during extraction. Polishing or rubbing the surface of linen during manufacture to make it smooth may weaken it structurally. Wool fibre damage may occur at a microscopic level during shearing, fulling and felting.

Damage to all types of fibres also occurs during use, with outward facing surfaces getting rubbed, flattened, cracked and broken. More obviously, damage is inflicted by sharp implements, and creasing too can inflict permanent damage at a molecular level. Linen fabrics become matted, and woollen threads pile and rub away; in each case the fibres weaken, thin and break.

Deterioration may be compounded by textiles taking up human chemicals that may cause odour, staining and weakening due to the bacteria living on and destroying the fibres. Leather may also become brittle and dark in colour because of the oxidation of preservatives, such as the oils and fats applied during the tanning process.[6] Washing in order to clean away these forms of dirt will also damage fibres. Cleaning agents, which are normally alkaline, attack fabrics, particularly those made from cellulose fibres, while the mechanical motion entangles and blends the roughened 'fibrils' (thread-like structures), causing material to mat.[7]

Noble metals, the category of metal to which silver and gold belong, are less subject to corrosion.[8] Indeed gold does not corrode at all, however, it is a soft metal and can be very weak and brittle, making it fragile and easily damaged.[9] Silver is second only to gold in purity, and does corrode. The corroded silver is called horn silver, and it occurs in two forms. If the corrosion process is slow and has only converted the silver in small quantities, a core of the original silver thread will be retained inside the protective horn silver crust; however, if the corrosion is deeply embedded and there is a thick and/or swollen crust of horn silver, no silver metal will have survived, leaving an empty cavity where it originally lay. Silver can also be brittle and extremely fragile and it must be handled with care so as not to shatter it.[10]

If silver-gilt threads are damaged through wear or use and the gold layer is removed or breached, the exposed silver beneath may begin to corrode in the same way as if the thread were solid silver. In such cases the thin gold covering can also become covered by

---

6 Cronyn, *Elements of Archaeological Conservation*, 15–16, 265–266, 286; Landi, *Textile Conservator's Manual*, 18–19.
7 B. Cooke and B. Lomas, 'Evidence of wear and damage in ancient textiles', in *Textiles in Northern Archaeology*, eds P. Walton and J.-P. Wild, 215–226 (218).
8 Noble metals are purer than other metals and occur naturally, not as an ore that needs to be refined.
9 Cronyn, *Elements of Archaeological Conservation*, 171, 235–237; Landi, *Textile Conservator's Manual*, 12; Pye, 'Conservation and Storage', 411.
10 Cronyn, *Elements of Archaeological Conservation*, 182, 230–232, 234; Pye, 'Conservation and Storage', 411.

corrosion. Silver-gilt threads should thus be treated as though they were pure silver, since the gold only superficially masks the core constituent. Such forms of deterioration can be an issue if an object is buried within an archaeological context, as we shall see below.

Survival in the archaeological record

Once embroidery has become part of the archaeological record, it is the interaction of the already deteriorating fibres with their surrounding environment that determines to what extent the textile will survive, if at all. Particular combinations of fibres, leather and metal, together with variable burial environments, lead to different levels of degradation. Textiles are generally found either in areas of extreme dryness, permafrost or water-logging; or carbonised between layers of burnt material; or in contact with corroding metals.[11] Embroideries found in archaeological contexts in the cohort covered by this book provide examples of finds from all environmental conditions except permafrost.

Dry environments

Desert regions provide excellent survival conditions for a wide range of fibres, especially woollen and silk textiles and embroideries.[12] Controlled environments in north-west Europe may give rise to similar conditions: for instance, hermetically sealed finds where the absence of oxygen provides particularly good conditions for embroidery's survival, since the activity of aerobic organisms on organic materials is prevented.[13] However, embroideries found with only the metal threads enduring indicate that conditions have been too acidic to allow for survival of protein fibres.[14]

Storage in a container provides a dry micro-climate once sealed. The most frequent examples are coffins. Although when the dead body is first interred in a casket there will be some moisture and fluid, this dries out over time leaving a stable microclimate. Embroideries placed in coffins tend to survive in fragmentary form: the bacteria active in the damp phase destroy certain parts, particularly those already in a fragile and fragmentary condition, leaving the remainder to survive once the space inside the casket dries out. As Cronyn has noted, this form of preservation is enhanced when the archaeological material is covered quickly and there is little further fluctuation in the surrounding environment.[15] Embroideries placed in the casket after the space inside has entirely dried out, for

---

11 J. Jones, J. Unruh, R. Knaller, I. Skals, L. Ræder Knudsen, E. Jordan-Fahrbach and L. Mumford, 'Guidelines for the Excavation of Archaeological Textiles', in *First Aid for the Excavation of Archaeological Textiles*, eds Gillis and Nosch, 5–30 (7). See also Cooke and Lomas, 'Evidence of wear and damage', 215–218.
12 Wild, *Textiles in Archaeology*, 7.
13 Cronyn, *Elements of Archaeological Conservation*, 24–25.
14 Acid destroys the hydrogen bonds in silk fibres. L.R. Sibley and K.A. Jakes, 'Survival of protein fibres in archaeological contexts', *Science and Archaeology*, 26 (1984), 17–27.
15 Cronyn, *Elements of Archaeological Conservation*, 28.

example, donations to a saint centuries after the death, where a stable microclimate has already been established, enable embroideries to survive in far better condition (for example, the stole, maniple and ribbons placed in the tomb of St Cuthbert).

Wet environments

Wet conditions encompass natural environments such as bogs and lakes, and man-made contexts such as burial mounds, and urban developments on ground where water does not drain well. Although the immersion of organic material such as textile or leather in water creates a chemical reaction which results in slow deterioration, organic material may be extremely well preserved due to a lack of oxygen and the penetration of humic, organic acids in the water – a good example being the Orkney hood.[16]

Fibres such as wool can survive in damp conditions as long as they are raised above the wet area. Such situations occur within burials where grave goods are stacked on top of one another (raising them above the damp line), and protected by surrounding materials such as metalwork. In this context, embroidery may survive because it has been kept hydrolysed by the moisture in the air.[17] This was the case for the embroidered seam discovered in the ship burial, Mound 1, at Sutton Hoo (Sutton Hoo A).

Similar conditions can also be found in urban settings. During the later part of the early medieval period, urbanisation compounded damp conditions in places such as Dublin, London and York, which created the ideal environment for the survival of protein-based materials such as silk, wool and leather. Damp soils, particularly those situated next to rivers, consist of typical wet burial conditions – damp and anaerobic – which, as Elizabeth Wincott Heckett has pointed out, are perfect for the preservation of protein fibres.[18] In London, the surviving protein materials indicate a lack of oxygen and high levels of organic compounds in the soil, which may have resulted from the spreading of midden waste. York also has high survival rates for both leather and protein-based textiles. The waterlogged site was found to contain extremely organic and anoxic (oxygen-depleted) deposits. Despite the fact that a sub-soil of sandy clay, laid over by a layer of loamy deposits, was not conducive for the

---

16 Cronyn, *Elements of Archaeological Conservation*, 26–27; J.P. Wild, *Textile Manufacture in the Northern Roman Provinces* (Cambridge: Cambridge University Press, 1970), 41–42; Wild, *Textiles in Archaeology*, 7–8.
17 This is how the embroidered seam and the now disintegrated embroidered shoes from the Ship Burial in Mound 1 at Sutton Hoo survived. See K. East, 'The Shoes', in *Sutton Hoo Ship-Burial volume 3*, eds R. Bruce-Mitford and A.C. Evans (London: British Museum Publications, 1983), 788–812 (788); Crowfoot, 'The Textiles', 411; R. Bruce-Mitford, *The Sutton Hoo Ship-Burial volume 1: excavations, background, the ship, dating and inventory* (London: British Museum Publications, 1975), 539–40.
18 E. Wincott Heckett, *Viking Age Headcoverings* (Dublin: Royal Irish Academy, 2003), 1.

survival of organic material, pits cut into the levels that lay on top of the deposits were filled with organic debris. Because later layers produced rich deposits of organic material, ideal conditions were produced for protein-based fibres to survive.[19]

Carbonisation and pseudomorphs[20]

The carbonisation of embroideries is rare and the extant number of carbonised textiles is small because they are usually destroyed by the fire. For a textile to survive it needs to go through the oxidation process slowly, under controlled conditions, resulting in the fibre structure being only partially rearranged in a transformation described as charring. Charred fibres emit no light, meaning the only remaining information they offer is a silhouette and indications of size. This makes the identification of fibre type extremely difficult.[21] The correct conditions are found at sites at which cremations have taken place. Such conflagrations take place under controlled conditions, leaving at least some of the fibre structure from silver thread undamaged at a microscopic level (see Ingleby). In other situations a textile may survive the burning process if a second phase of different conditions inhibits the chemical change initiated by burning. The Llangorse textile, which was burnt and then submerged in Llangorse Lake, is an example of this.[22] The fibres underwent two forms of preservation: by fire and then by water. Oxidation would have altered the chemical properties of the cellulose linen fibre and the protein-based silk threads, making them susceptible to preservation in water.[23] This part of the lake probably contained a large quantity of organic material in its silts due to the presence of a wooden palace which had been built on a crannóg in the lake, giving the water similar preserving qualities to a bog.[24]

---

19 R.A. Hall, *Anglo-Scandinavian Occupation at 16-22 Coppergate: defining a townscape*, The Archaeology of York, Anglo-Scandinavian York 8/5 (York: Council for British Archaeology, 2014), 531–536; J.A. Spriggs, 'Conservation of the Leatherwork', in *Leather and Leatherworking in Anglo-Scandinavian and Medieval York*, The Archaeology of York The Small Finds 17/16, eds Q. Mould, I. Carlisle and E. Cameron (York: Council for British Archaeology, 2003), 3213–3221 (3213).
20 No negative casts of embroidery from this period have been discovered so this section will only be discussing pseudomorphs.
21 Cooke, 'Fibre Damage', 9; M. Goodway, 'Fiber Identification in Practice', *Journal of the American Institute for Conservation*, 26 (1987), 27–44 (28).
22 Louise Mumford, the textile's curator, believes the textile was purposely dropped into the water because of the way it was folded and the location of the lake silts when the piece was found. Pers. comm. Louise Mumford (2015).
23 L. Mumford, 'The Conservation of the Llangorse Textile', in *Proceedings of the 8th ICOM Group on Wet Organic Archaeological Materials Conference: Stockholm 2001*, eds P. Hoffmann, J.A. Spriggs, T. Grant, C. Cook and A. Recht (Bremerhaven: International Council of Museums (ICOM), 2002), 471–491 (471, 473); Pers. comm. Louise Mumford (2015).
24 E. Campbell and A. Lane, 'Llangorse: a 10th-century royal crannog in Wales', *Antiquity*, 63 (1989), 675–681 (the full publication is forthcoming as we go to press).

Finally, a pseudomorph is a positive cast that is created when a textile which has been lying next to a metal object in neutral or alkaline oxygenated surroundings, absorbs the insoluble products, sulphides or salts that the metal exudes as it corrodes. During this transformation the organic fibres are destroyed in a process called mineralisation and the corrosive products take on the structure and form of the textile, creating a positive cast. Iron provides a good example because iron oxide changes the organic matrix of the textile into an iron compound that goes on to survive well in the archaeological record.[25] Within the confines of this project, the term pseudomorph is used to describe the complete replacement of fibres with metal.[26] The only examples discussed in this book are the sleeve cuff from Mound 14, Sutton Hoo (Sutton Hoo B) and a garment of embroidery found in contact with the blade of a knife from Worthy Park.

Excavation of embroideries

The survival of an embroidery in the archaeological record means a level of balance has been created between the artefact and the surrounding environment. As soon as a textile is lifted from such a setting, the equilibrium between the chemical, biological and support structures is destroyed. As a result the fibres begin to deteriorate at a more rapid rate, meaning the textile needs to be conserved swiftly and appropriately in order to recreate the balance and slow down degradation.[27] This results in the need for continued curatorial intervention beyond the original archaeological

---

25 Cronyn, *Elements of Archaeological Conservation*, 172; Wild, *Textile Manufacture*, 43; Wild, *Textiles in Archaeology*, 9–11.

26 R. Gillard and his co-authors have stated that a pseudomorph is the complete replacement of the textile with a non-organic matrix while the term 'mineralisation' can refer to a partial or complete replacement: K. Anheuser and M. Roumeliotou, 'Characterisation of Mineralised Archaeological Textile Fibres through Chemical Staining', *The Conservator*, 27 (2003), 23–33 (23); R.D. Gillard, S.M. Hardman, R.G. Thomas and D.E. Watkinson, 'The Mineralization of Fibres in Burial Environments', *Studies in Conservation*, 39 (1994), 132–140 (132). There are no examples of partially replaced embroideries; however, for future reference, any partially replaced textiles or embroideries will be given the term mineralised.

27 Cronyn, *Elements of Archaeological Conservation*, 29–30; Pye, 'Conservation and Storage', 392; Jones et al., 'Guidelines', 10–11.

28 Good working practice now dictates integrated planning that organises every aspect of a dig from pre-assessment through to post-excavation work and publication. Where it is known that textiles might be recovered, access to a trained curator of textiles to advise, plan and work with the excavation team and clean, stabilise and record the textiles, and contribute a specialist report, should be incorporated into the plan. For further information, see Anon, *Standard and Guidance: Archaeological Excavation* (Reading: The Chartered Institute for Archaeologists, 2014), and Jones et al., 'Guidelines', 5–6. These publications also give advice on how to proceed in situations in which textiles are a non-anticipated find.

environment.²⁸ Once the textile is at the laboratory, the conservator can begin to assess how it should be cleaned and stabilised. The better informed the conservator is as to the conditions under which a textile was found, the better prepared he or she will be as to how to proceed.

Initial data

A grounding in the difficulties associated with early medieval embroidery finds provides a platform for initial discussion of the technical data gathered together in Tables 1A–G. The level of information we can derive from the data gathered from this survey is limited by the small size of the dataset and lack of certainty with regard to certain aspects of the finds, nevertheless, a number of preliminary observations can be made.

Dates (Fig. 19)

Although Diagram 1 shows that there is a spread of surviving embroideries right across our timeframe, the majority are found in 8th- to 9th-century contexts. This is the period into which the 12 wool and silk fragments from the Oseberg ship burial

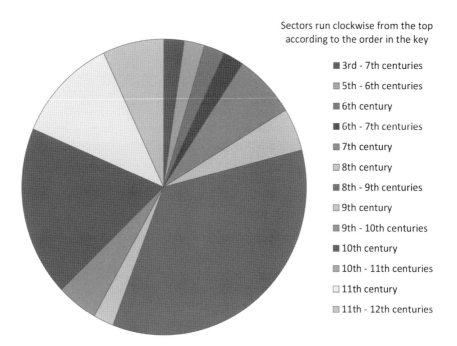

Figure 19. Diagram 1: Distribution of surviving embroideries by period. Sectors that encompass more than one century include embroideries where there is insufficient evidence to attribute the work to a single specific century

fall. The hermetically sealed inhumation provided good conditions for survival of the textiles, particularly those made from protein fibres, and despite it having been robbed at some stage in its history.[29] The three sets of embroideries now housed in the treasury of St Catherine's Church in Maaseik, Belgium also date to this period. These have survived in good condition because their history indicates that they were treated as contact relics associated with the two sister saints, Harlindis and Relindis, who founded the abbey church at Aldeneik, where the embroideries were originally kept.[30] The embroideries were stored in a reliquary, which aided their continued survival.

The second largest surviving cluster of embroideries dates to the 10th century and consists of four pieces from two churches and four from archaeological contexts. Three of the four textiles from religious settings were discovered in the coffin of St Cuthbert at Durham Cathedral, England (Durham C–E), while the fourth was found in the tomb of St Ambrose in the Basilica Ambrosiana in Milan, Italy. Once again, these items had good conditions in which to survive, having been placed inside almost hermetically sealed coffins after the bodies of the deceased had 'settled' or decomposed, meaning that many of the bodily fluids that could have damaged the embroideries had already seeped to the bottom, away from where the textiles were laid.

The four pieces discovered in archaeological contexts are good examples of what can endure when the depository conditions are conducive to the survival of both cellulose and protein fibres, and the excavators have knowledge of what to look for. The first is a section of wool embroidery from Dublin, Ireland. It was found during the 1974–1981 digs that took place in Wood Quay and Fishbourne Street. In Fishbourne Street a cess pit was excavated and it was in this that the embroidery, along with many other textiles, was discovered.[31] The second is a vamp from a shoe (that is, the top of the shoe), unearthed during the 1961–1971 excavations at Castle Yard, Street 2, Winchester. The surviving section of the vamp (see Glossary) is decorated with rows of stitching that J.H. Thornton thought probably formed a vamp stripe.[32] The third and fourth objects from this 10th-century group were both discovered in York. The first of these two items is a silk pouch that was found in a probable workshop setting at 16–22 Coppergate (York A). The bag itself had survived particularly well, although the embroidery was somewhat disintegrated. The final item is a decoratively bound edge, probably the cuff and lower portion of a sleeve (York B). This was excavated

---

29 Due to the fragmentary nature of the pieces it is not known whether they came from multiple textiles or whether some pieces come from a single piece. They have therefore been catalogued individually.

30 A contact relic (*brandea*) is one that has been touched to the body of the saint but is not part of the saint's corporal body. See J. Crook, *English Medieval Shrines* (Woodbridge: Boydell, 2011), 16–18, C. Freeman, *Holy Bones, Holy Dust: How Relics Shaped the History of Medieval Europe* (London: Yale University Press, 2011), 26–27.

31 See F. Pritchard, 'Textiles from Dublin', in *Vikingatidens Kvinnor*, eds N. Coleman and N. Løkka (Oslo: Scandinavian Academic Press, 2014), 225–240.

32 J.H. Thornton, 'Shoes, Boots, and Shoe Repairs', in *Object and Economy in Medieval Winchester*, ed. M. Biddle, Winchester Studies, 7.ii (Oxford: Clarendon Press, 1990), 591–617 (594, 596).

at High Ousegate, a street thought by scholars to have been located on the edge of the craft-making district, of which 16–22 Coppergate was also a part.[33] All these archaeological sites had moist conditions.

While these clusters mean that some periods furnish us with a high number of extant embroideries, it is notable that across many of the other century or century-to-century intervals, only a single piece survives. Thus the five equally sized smallest segments in the pie chart each equate to a single embroidery sample per segment, while two more segments contain two apiece.

Object type (Fig. 20)

Table 1B shows the twenty different types of object, or parts of an object, on which embroidery has been discovered. In some cases, there is some doubt as to

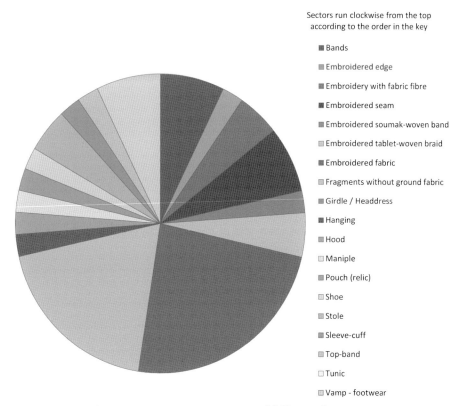

Figure 20. Diagram 2: Proportions of different object types

---

33 R.A. Hall, D.W. Rollason, M. Blackburn, D.N. Parsons, G. Fellows-Jensen, A.R. Hall, H.K. Kenward, T.P. O'Connor, D. Tweddle, A.J. Mainman and N.S.H. Rogers, *Aspects of Anglo-Scandinavian York*, The Archaeology of York, Anglo-Scandinavian York 8/4 (York: Council for British Archaeology, 2004), 475, 478, fig. 128.

what the embroidery originally was and these pieces have been subdivided into broad categories, the largest of which, as Diagram 2 shows, is embroidered (but unidentified) fabric, for which there are twenty extant examples ('Embroidered fabric'). Many textiles and their embroideries decay and become fragmented over time once buried or discarded. Some lose further fibres once excavated, especially if they are not stabilised and/or stored correctly. The second largest group, which consists of eight examples, is fragments of embroidery without any discernible ground fabric ('Fragments with no ground fabric'). All these pieces come from the Oseberg Ship Burial (Oseberg E–L). Two further examples consist of embroidery that displayed traces of fabric fibres but not a complete ground fabric ('Embroidery with fabric fibre'). The first of these is the pulled wire embroidery from Ingleby. In 1955 or 1956 this piece was analysed by the Shirley Institute where it was discovered that possible carbonised material was sandwiched between the stitches at the front and the carrying stitches at the back of the embroidery, indicating that the pulled wire had been worked through a ground fabric, the fibres of which had subsequently been sandwiched between the stitches on the front and the carrying threads to the back.[34] Such incidences tend to occur when the conditions of burial are conducive to the survival of one or more types of fibre, either protein, cellulose or metal, but not others, resulting in partial, or almost total destruction of the ground. The second piece that falls into the category 'Embroidery with fabric fibre' is a gold-work embroidery from Milan. In the 11th century the embroidery was patched and re-lined and in 1863 the then Prevosto, Monsignor Rossi, had the embroidery remounted onto a new ground fabric. During a 1940s conservation programme the embroidery was remounted once again on to a new silk ground fabric and at this time fibres of violet and black silk from what are assumed to be the original ground fabric were discovered.[35] At some point in its history the embroidery has been cut up into its individual design elements, and the original fibres can be seen around the edges of a number of these sections (Pl. 12a).

The three surviving 'Embroidered seams' are notable for combining functional stitching with decoration. The first and second example are the seams from Sutton Hoo (Sutton Hoo A) and Dublin. Sutton Hoo A is a possible cushion or pillow that survives as two pieces of fabric that have been sewn together using a decorative looped stitch (see Pl. 2a & Fig. 1) instead of a purely functional stitch that could have remained invisible to the viewer. The piece from Dublin is fragmentary and its original function

---

34 Unpublished letter from 22 February 1957, from English Heritage Archive. I am grateful to Claire Tsang and Kirsty Stonell Walker, Archive and Information Team at English Heritage, Fort Cumberland, for a copy of this letter.
35 See G.M. Crowfoot, 'Note on a Fragment from Basilica Ambrosiana in Milan', in *The Relics of St Cuthbert*, ed. C.F. Battiscombe (Oxford: Oxford University Press, 1956), 392; A. de Capitani d'Arzago, *Antichi della Basilica Ambrosiana*, new series 2 (Milan: Biblioteca de L'Arte, 1941), 41; H. Granger-Taylor, 'The Two Dalmatics of Saint Ambrose?', *Bulletin de Liaison de Centre International D'Ètude des Textiles Anciens*, 57–58 (1983), 127–173 (129).

is unknown. The third example is the seam that joins the front and back pieces of a tablet-woven band together, forming a *bursa* (Utrecht). This seam was stitched using a functional stitch, buttonhole stitch (see Glossary), decoratively. The ornamental effect was further enhanced by the use of a coloured embroidery thread that would have stood out against the colour of the tablet-woven band.

'Tablet-woven bands' (see Glossary) are worked as independent textiles and come in many widths and lengths. Once finished they can be attached to other fabrics to create decorative borders, for example, on the hems or edges of clothing. They can also be folded and stitched to create a pouch or small bag, such as that from Utrecht. Embroidery worked on tablet-woven bands appears on two objects from either end of our timeframe. The earliest comes from Mitchell's Hill and is a fragment of possible embroidery attached to a tablet-woven band discovered lying between a textile and one of a set of wrist-clasps that are thought to date to the 6th century (see Fig. 8).[36] Since this piece is missing from the Ashmolean Museum in Oxford where it was originally stored, it is not possible to confirm whether the buttonhole stitch is actually embroidery – functional stitching worked decoratively – or entirely utilitarian. The second example was found in the coffin of St Cuthbert in Durham Cathedral (Durham B) and probably dates to *c.* late 8th century. There is a query regarding this example as well. Although previous researchers have concluded that the stitching was buttonhole stitch,[37] the microscopic images and analysis undertaken for this research show that it could actually be satin stitch or a form of wrapping stitch, but this cannot be confirmed without being able to view the reverse of the embroidery clearly (Pl. 12b, c).

The two stoles (a form of liturgical vestment, see Glossary) are finds from tombs. The first from the tomb of St Cuthbert (Durham D) dates to *c.* 909–916. The second was discovered in a tomb ascribed to Bishop William de Blois (d. 1236), in the Lady Chapel at Worcester Cathedral.[38] All the other object types survive as single examples. They represent a fairly broad range of secular dress and soft furnishings of the timeframe, including pieces from elite circles: a possible girdle or headdress part (Durham C), a tunic (Llangorse) and a hanging (Bayeux). There are items that may have been made and/or used by those of middling rank too: a hood (Orkney), a pouch (York A), an embroidered vamp from a shoe (Coventry), an embroidered top-band, also relating

---

36 G.M. Crowfoot, 'Anglo-Saxon Tablet Weaving', *The Antiquaries Journal*, 32, 189–191 (189).
37 G.M. Crowfoot, 'The Braids', in *Relics of St Cuthbert*, ed. Battiscombe, 433–469 (462); H. Granger-Taylor, 'The Weft-Patterned Silks and their Braid: the remains of an Anglo-Saxon dalmatic of *c.* 800?', in *St. Cuthbert and his Cult and his Community to AD 1200*, eds G. Bonner, D. Rollason and C. Stancliffe (Woodbrige: Boydell Press, 2002), 303–327 (324).
38 W.H. St John Hope, *Proceedings of the Society of Antiquaries*, 2nd ser., 14 (1892), 196–200 (197); M. Perkins, 'A Stole, Maniple and Four other Embroideries in Worcester Cathedral Library', *Archaeology at Worcester Cathedral: report of the sixteenth annual symposium March 2006* (2007), 3–12 (4); C. Wild, *An Illustration of the Architecture and Sculpture of the Cathedral Church of Worcester* (London: Charles Wild, 1823), 20.

to footwear (London A) and an embroidered edge, probably the edge of a sleeve cuff (York B). There is also a further example of an ecclesiastical vestment, a maniple (Durham E). However, as might be expected, there is little indication of embroidered objects that may have been used by those of lower rank. Such individuals would have had fewer textile objects available for decoration and what they did own would have been continually recycled until it disintegrated.

The materials (Fig. 21)

Some interesting observations can be made with regard to materials, although the dataset is small, and possibly distorted by the spike in numbers generated by the Oseberg embroideries. What is of particular interest here is the interplay between different types of ground fabric and embroidery thread. In some cases, however, poor condition of surviving materials precludes useful interpretation. Diagram 3 shows that the majority of extant pieces (26) are fragmentary, but two are complete, and a substantial number (11) – are partially complete. The rest are either mineralised or replaced (Alfriston, Sutton Hoo B, Worthy Park). Mineralisation or replacement of fibres may create detailed negative copies of the original textiles, enabling the type of fibre used in their creation to be determined. Following scientific analysis,

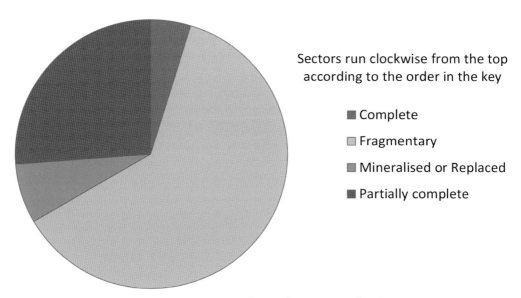

Figure 21. Diagram 3: Condition of surviving embroideries

---

39 P. Walton Rogers, 'The Textiles from Mounds 5, 7, 14 and 17', in *Sutton Hoo: an Anglo-Saxon princely burial ground and its context*, ed. M. Carver (London: British Museum Press, 2005), 262–268 (214).

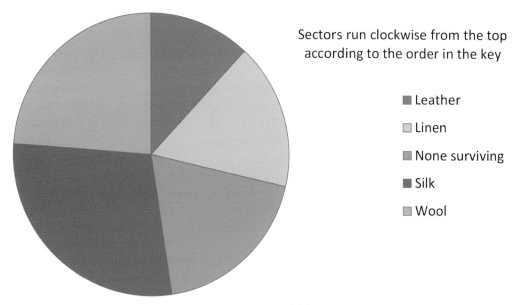

Figure 22. Diagram 4: Ground fabric materials

Penelope Walton Rogers was able to state that the ground fabric of Sutton Hoo B had been linen woven in a fine tabby weave of 30 × 28 threads per cm.[39] On the other hand, descriptions of replaced embroidery produced before such advanced analysis was available cannot be relied on for accuracy. A.F. Griffith and L.F. Salzmann describe the replaced ground of the Alfriston embroidery as linen in their 1914 publication, but it is not clear whether they actually knew that it was linen or if they used the term more generally.[40] For Worthy Park, although materials could not be determined, Elisabeth Crowfoot was able to go so far as distinguishing the type of weave as tabby, and the thread count as 22 × 22 threads per cm.[41]

Extant embroidery has been found on four different types of ground fabric: leather, linen, silk and wool. Diagram 4 (Fig. 22) shows that survival rates across differing fibres are roughly comparable. Silk yields the greatest number of examples, at 12, wool next, with 10, while linen has seven and leather five. As already discussed above, those with no surviving ground fabric are the Oseberg pieces. The greater incidence of surviving silk pieces is not an indication that it is the most robust fibre, it relates to discovery contexts. Most embroideries worked on a silk ground fabric have been found in religious settings, whether in a church building itself or in the coffin of a saint or member of the clergy, buried within a cathedral. These contexts provide

---

40 A.F. Griffith and L.F. Salzmann, 'An Anglo-Saxon Cemetery at Alfriston, Sussex', *Sussex Archaeological Collections, Relating to the History and Antiquities of the County*, 56 (1914), 16–51 (34).
41 Crowfoot, 'Chapter 5: The Textile Remains', 192.

good survival conditions. Silk was a precious and much sought-after textile/thread and difficult to obtain (for further discussion, see p. 114). The fibre was utilised by the wealthy and influential to demonstrate the power of the Church. It is therefore hardly surprising that embroidery worked on silk ground fabric, regarded as both precious in terms of its exchange value and its religious symbolism, has survived in the largest quantities.

The second largest category of extant ground fabric is wool. Most of these examples have been found in burials, with two being discovered in probable female inhumations (Kempston and Mitchell's Hill). The example from Kempston was found in a small copper alloy box sealed with a close fitting lid, giving the wool embroidery the perfect conditions in which to survive. Four pieces – the majority – come from the Oseberg Ship burial, which was hermetically sealed and conducive to the survival of protein fibres. Another ground fabric was found in a pile of wool textiles in the ship burial of Mound 1 at Sutton Hoo (Sutton Hoo A). This find was in a protected position underneath a silver 'Anastasius' dish and raised above the base of the burial chamber. Other surviving pieces were discovered in waterlogged or damp conditions that create an equilibrium in which the wool stays moist but does not rot. These were a bog (Orkney hood), a cesspit (Dublin) and the grounds of a possible workshop (York B).

Examples of linen date from across the timeframe: the 7th-century cuff, Sutton Hoo B; the 8th- to 9th-century embroideries from Maaseik; the 9th- to 10th-century tunic from Llangorse and the 11th-century hanging now housed at Bayeux.[42] These all involve contexts associated with the elite of society. This suggests that linen was favoured by those in higher ranking circles and, because of the extensive processes and expense involved in its production, and the amount of man- and woman-power required, this would restrict linen to those with the means to buy or manufacture it. On the other hand, Walton Rogers has pointed out that surviving examples of linen from this period are numerous and it should not be considered a luxury in the way silk was.[43]

All surviving examples of embroidered leather are associated with footwear. The earliest piece is a stitched vamp from Winchester which dates to the early 10th century. A second group consists of an embroidered top-band and two vamps from London. The top-band dates to between 1050 and 1100 while the vamps both date to between c. 1070s and 1080s. There is also is a near complete shoe with an embroidered stripe running down the vamp from Coventry dating to the late-11th to early-12th centuries.[44]

Examination of ground fabric weaves reveals that it is overwhelmingly tabby weave that survives. Tabby weave is the most basic type of woven textile where the

---

42 The Alfriston piece has also been included in this group, although we do not know for certain whether the ground fabric was linen (see earlier discussion).
43 P. Walton Rogers, *Cloth and Clothing in Early Anglo-Saxon England: AD 450–700* (York: Council for British Archaeology, 2007), 235.
44 My thanks go to Frances Pritchard for dating the Coventry shoe.

weft threads are passed alternately under and over the warp threads (see Glossary). It is logical to infer that this weave was utilised most often and as a result, survives in greatest quantities. In their work on clothing and textiles, Penelope Walton Rogers and others have demonstrated incidence of tabby-woven fabrics across the social spectrum.[45] However, Table 1D shows that of the surviving *embroideries*, tabby ground fabrics are more commonly associated with elite examples: the Bayeux Tapestry, Durham A–F, Llangorse, Maaseik A–C, Oseberg A–D, Sutton Hoo B and Worcester. Only Worthy Park provides a possible example of an item that may have been used by someone of lower rank. This suggestion is based on archaeological evidence from the burial in which the embroidery was found, which indicates that the deceased was not of high rank but had been a member of the 'ordinary rural [early] Anglo-Saxon settlement' to which the cemetery is thought to have belonged.[46] However, Elisabeth Crowfoot's thorough analysis suggests it could equally have been a piece of fine textile and belonged to someone of rank.

The data is more patchy for the remainder of the ground fabrics. The five leather pieces are of course excluded here because they are not woven textiles. The weave of seven further examples is unknown. Four more are tablet-woven bands, two of silk (Durham B and Utrecht), and two of wool (Mitchell's Hill and Orkney). Two embroideries dating to the 7th century are stitched onto a broken diamond twill wool (Kempston and Sutton Hoo A). There is one example of a compound twill fabric, which is woven from silk (York A).[47]

Embroidery thread (Fig. 23)

The extant embroideries were worked in six types of embroidery thread: gold, silver and silver-gilt metal threads; linen; silk and wool (see Table 1E and Diagram 5). There is a seventh category, 'replaced' (where the thread has been wholly/partially replaced, usually by corrosive metal products while buried). This is a found classification, rather than a worked category, but it has been included here because the thread type can still be identified. The majority of surviving threads are silk (23). As we have seen, the largest category of surviving ground fabrics is also silk, and silk thread and silk ground fabric commonly appear together. However, there is also a high percentage of silk embroideries with no surviving ground fabric. These are the eight examples

---

[45] Walton Rogers, *Cloth and Clothing*, 235; G.R. Owen-Crocker, *Dress in Anglo-Saxon England, revised and enlarged* (Woodbridge: Boydell Press, 2004), 291–293.
[46] For discussion on the possible status of those buried in the cemetery see S. Chadwick Hawkes, 'Introduction', in *The Anglo-Saxon Cemetery at Worthy Park, Kingsworthy, near Winchester, Hampshire*, ed. S. Chadwick Hawkes with G. Grainger, Oxford University School of Archaeology Monograph 59 (Oxford: Oxford University School of Archaeology, 2003), 1–11.
[47] For detailed analyses of the implications of materials and weaves and their place in early medieval society see the bibliography for relevant publications G. and E. Crowfoot, P. Walton Rogers, F. Pritchard and H. Granger-Taylor.

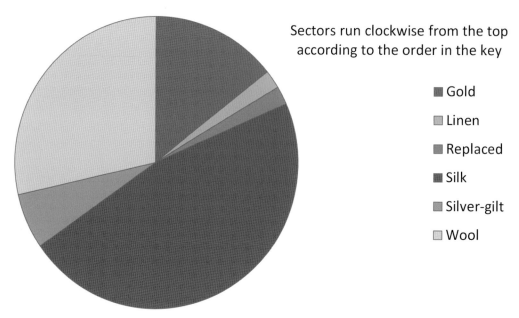

Figure 23. Diagram 5: Categories of surviving embroidery threads

found in the Oseberg Ship Burial (Oseberg E–L). Arne Emil Christensen and Margereta Nöckert have suggested that the fineness of the surviving stitches of these embroideries indicates that the ground fabric on which they were worked was originally linen.[48] This would lead to the hypothesis that the burial conditions of the ship were good for the survival of protein fibres but not cellulose fibres. There is further surviving evidence for the practice of combining linen ground fabric with silk embroidery thread in the Llangorse and Maaseik A–C examples.

There are 14 examples of wool thread. Again, there is a correlation between the incidence of wool yarn and wool ground fabric, for which there are 10 objects: Dublin, Kempston, Mitchell's Hill, Orkney, Oseberg A–D, Sutton Hoo A and York B. The other four pieces involve two examples of wool embroidered leather (Coventry and London A), and two examples of wool embroidered linen (Bayeux and Sutton Hoo B).

Gold thread also shows a good survival rate with seven extant examples: three on linen (Maaseik A–C) and four on silk (Durham C–E, and Milan). All of these pieces also include silk thread in the design, and, in all cases, as well as being used to embroider discrete elements or motifs, the silk thread has been used to couch the gold thread in place. Again the incidence of linen ground fabric worked with rich thread materials affirms its status.

---

48 A.E. Christensen and M. Nöket, eds, *Osebergfunnet: Bind IV Tekstilene* (Oslo: Museum of Cultural History, 2006), 399.

There are three silver-gilt pieces: Durham F–G and Worcester. All of the embroideries which utilised this thread date to the end of our timeframe and can be seen as part of a move away from the use of gold thread to cheaper gilded silver. All three pieces were worked on a silk ground fabric fitting with the development of embroidery from the early medieval styles to those of the later *Opus Anglicanum* (see Worcester in the catalogue).

The other three categories, linen, silver and replaced threads, all survive as single extant examples. The linen thread occurs in only a very few places on the Bayeux Tapestry; the rest is embroidered in wool on a linen ground fabric. The example of silver 'thread' comes from Ingleby and is pulled wire as opposed to a gold or silver-gilt thread (see Glossary). The information for the replaced example discovered at Alfriston is sparse and what we have cannot be taken as reliable, as discussed above.

Where threads have been measured for thickness, the majority have been described by archaeological textile experts as fine (17 examples) or very fine (17 examples), with only two classified as medium (see Table 1E). Despite the small dataset, it does seem to indicate a preference for the utilisation of fine or very fine threads on embroidery right across the timeframe, particularly with regard to those examples circulating within elite or ecclesiastical settings. Conversely, since the Milan embroidery is one of the two pieces that makes use of medium threads, use of coarser threads appears not to have been confined to less affluent circles as one might suppose. The Milan embroidery is a 10th-century piece worked in silk and gold threads on a silk ground fabric and was discovered in the tomb of St Ambrose in the Basilica Ambrosiana, Milan, Italy. It was a piece made for someone with the means and power to commission workers to produce an embroidery in precious materials, which was then at some point later in its life donated to St Ambrose and thus removed from circulation in the secular world.

Stitches (Fig. 24)

The data on stitch types (Table 1F and Diagram 6) highlights that stem stitch is the most commonly used stitch across the cohort and timeframe, with 25 extant examples. The rest of the stitches can be divided into two groups with the first containing other commonly occurring stitches: split stitch (11), satin stitch (10) and couched work (10) (for illustrations of all stitches see Glossary). The second group consists of stitches with five or fewer extant examples: running stitch, underside couching and laid-work appear in two instances; buttonhole and tunnel stitch have four examples each, and chain and looped stitch have five, while counted work and raised plait have one example apiece. There is also an example where the stitch cannot be determined, Oseberg L.

Table 1F shows that many of the stitch types appear at different points in the timeframe, possibly suggesting particular stitches may have fallen out of favour, only to be brought back into use in later centuries, perhaps through local knowledge, when there was a need, or perhaps due to the influence of an external source creating a new fashion in the wake of movement of people, goods and ideas. However, there are

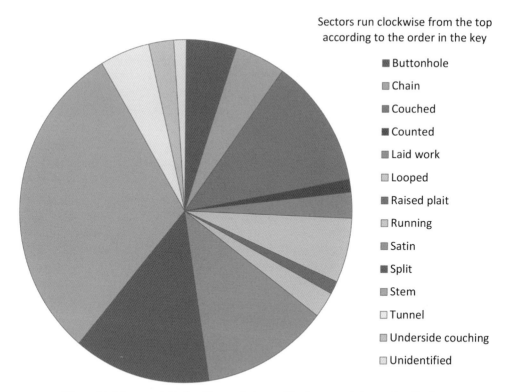

Figure 24. Diagram 6: Embroidery stitches utilised on surviving embroideries

also some stitches that appear to have been used on an ongoing basis. This applies to stem stitch in particular, which first appears in the archaeological record in the 6th to 7th centuries at Worthy Park, and for which, with the exception of the 8th century, there is always at least one surviving example for the period. The evidence is dominated in the 8th to 9th centuries by the nine fragments from the Oseberg Ship Burial (Oseberg A, D–K) and the sets of bands and monograms from Maaseik. Other stitches that express similar trends, although there are fewer examples, are split stitch and couched work. As I will go on to discuss in more detail below (see Chapter 4), we can hypothesise that the adoption of Christianity and the need for elaborate vestments helped establish and cement the use of these stitches across the period.

Looped stitch appears first on the Orkney hood, dating from between the 3rd and 7th centuries, and then in the early 7th-century burial at Sutton Hoo (Sutton Hoo A). It then disappears from the archaeological record to re-emerge in the 9th to 10th century at Ingleby, and then twice more in the 10th century in Dublin and York. This may be because looped stitch was primarily a Germanic and then a Scandinavian stitch that fell out of favour when the Anglo-Saxons converted to Christianity and turned their attention away from the northern pagan culture towards Christian Rome and Byzantium.[49] The stitch only reappears in the record in association with Scandinavian warriors (Ingleby) and when Viking trading hubs were established at York and Dublin. As archaeological evidence from the Netherlands and Birka in Sweden attest, the stitch continued to be worked on the Continent, and may thus have been re-introduced into England and Ireland through Scandinavian controlled territories. There is no known surviving evidence for looped stitch being used outside of these areas.

Chain stitch may be an example of a stitch that continued to be used but for which we do not have the surviving evidence. The stitch first appears on the Orkney hood. It then disappears from the record, only to re-emerge again in the 10th century on the relic pouch from York (York A), and again on the 11th-century Bayeux Tapestry. Since the surviving examples of this stitch appear in first Germanic and then Scandinavian influenced areas, it could be that this stitch had a similar life cycle to looped stitch.

Finally, there is one stitch that does not appear at all until the end of the period in the 11th century: underside couching, which most likely developed out of earlier forms of couched work at a time when couched gold-work survives to a lesser extent (see also discussion, pp. 134–135).

Not all the embroideries were worked in a single stitch; at least 21 pieces are examples of stitch combinations.[50] Such combinations appear across our timeframe but once again there is a spike between the 8th and 9th centuries owing to the groups of embroideries from Maaseik and Oseberg (Oseberg E–J). Particular stitch combinations may be attributed to the type of design, the materials used and the function of the finished piece. Some stitches are better suited to particular sizes and shape of area to be filled. Stitches can also be used to create varied effects within a design, for example, stem stitch worked in wool was used to create the striking circular design on the armour of the soldiers of the Bayeux Tapestry, but the same stitch is worked in much finer silk threads placed extremely close together in small areas of the ecclesiastical vestments Durham C–E. In these instances the effect the stitch produces is of fine opulence. Working stitches closely together also ensured the robustness of both the embroidery and the finished vestment, a useful attribute for items that were to be worn repeatedly. In this way, the use of

---

49 See A. Lester-Makin, 'Looped Stitch: the travels and development of an embroidery stitch', in *The Daily Lives of the Anglo-Saxons*, eds C. Biggam, C. Hough and D. Izdebska, Essays in Anglo-Saxon Studies 8 (Tempe Arizona: Arizona Center for Medieval and Renaissance Studies, 2017), 119–136.

50 As Table 1F indicates, there are eight pieces where the stitch types are unconfirmed. The other four pieces that have a question mark next to one stitch have at least one stitch type that is confirmed.

Design elements

Finally, we can analyse the data in terms of representation of particular design elements: that is, the range and occurrence of pictorial and geometric motifs deployed across the surviving embroideries. There are 27 different elements that occur singly or in combinations across the textiles (Table 1G) and four particular areas of interest are discernible. The first is the proportion of different design elements that occur across the surviving embroideries. The data has been set out across three pie charts in order to better encompass the large number of design elements. Diagram 7 sets out the surviving portions of embroidery design associated with the natural world, while Diagram 8 does the same for those that represent composite and symbolic elements drawn from the cultural environment. Diagram 9 shows the relative proportions of surviving embroideries showing simple repetitive patterns, or no patterns.

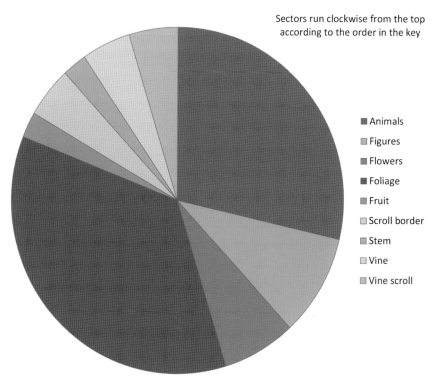

Figure 25. Diagram 7: Relative proportions of surviving design elements drawn from the natural world

Diagram 7 (Fig. 25) shows that two particular design elements representing the natural world have survived in the greatest quantities: foliage with 15 and animals with 12 examples. Indeed, when their numbers are compared across all the design elements, they have the highest and second highest overall survival rates. Foliage elements dominate, being a popular design throughout the timeframe and particularly from the 8th century onwards, with six examples from the 8th–9th centuries, four from the 10th and three from the 11th century. The number of animal elements is spread more evenly across the timeframe, appearing as early as the 7th century on the Kempston fragment through to the Bayeux Tapestry and a fragment from Durham (G) in the 11th century. The largest number date to between the 8th and 9th centuries and these are stitched on the embroideries from Maaseik (A and B) and Oseberg (A, E, F and G). Of the rest of the design elements to represent the natural world, all survive on one or two pieces except figures which appear on four (Bayeux, Durham D and E, Worcester) and flowers which occur on three (Bayeux, Durham C, Oseberg F). From this distribution we can cautiously deduce that foliage and animal design elements were the most commonly used.

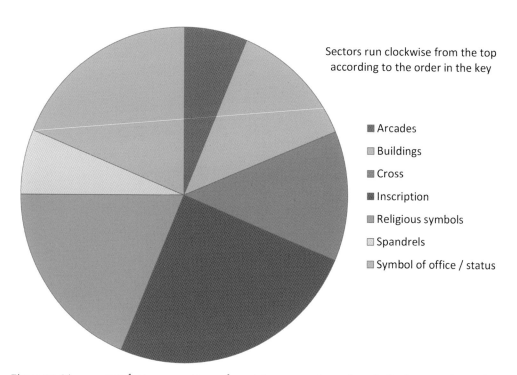

Figure 26. Diagram 8: Relative proportions of surviving composite and symbolic elements drawn from the cultural environment

Diagram 8 (Fig. 26) shows that the distribution of surviving composite and symbolic elements drawn from the cultural environment is more even: arcades and spandrels survive on one embroidery (Maaseik A), buildings and crosses on two embroideries each (Bayeux and Worcester, and Oseberg G and York A respectively) and religious symbols and symbols of office or statues appearing on three each (Bayeux, Durham D and E, and Durham D and E and Worcester respectively). Inscriptions survive on four embroideries (Bayeux, Durham D and E and Worcester). All of these design elements date from the late 8th–9th century onwards and all have religious connections, whether through design as in the case of the cross found in the Oseberg ship burial, or because they were made for or given to an ecclesiastical institution at some point in their biographies. With the exception of Oseberg G it is because these embroideries were associated with religion that they have survived.

Diagram 9 (Fig. 27) shows the relative proportions of surviving embroideries showing simple repetitive pattern or no pattern. Of the design elements that are included in Diagram 9, geometric elements and bars survive in the greatest numbers with eight examples. Geometric design elements first appear on embroideries dating

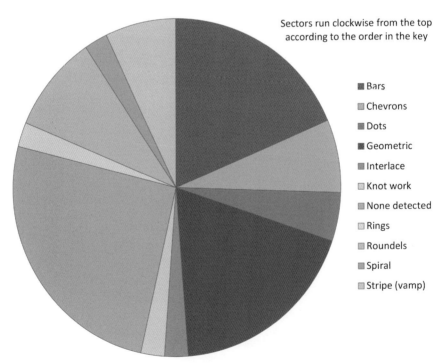

Figure 27. Diagram 9: relative proportions of surviving embroideries showing simple repetitive pattern, or no pattern

to between the 5th and 6th centuries. They then disappear from the record until the 8th–9th century and continue up to the 10th. Bars appear in the early period between the 7th–9th centuries and then disappear until the 11th–12th centuries. Such distinct clusters of design utilisation cannot be explained by type of embroidery because both design types appear on both religious and secular pieces. It is probable that these designs, like certain stitches, waxed and waned in popularity. The rest of the purely patterned design elements that make up Diagram 9 survive in smaller numbers, appearing rarely across the timeframe.

Of the embroideries that show no pattern, six can best be understood as showing the embroiderer expressing herself artistically at the same time as fulfilling a functional need. In other words, the embroidery on the pieces from Dublin, London A, Mitchell's Hill, Orkney, Sutton Hoo A, Utrecht and York B secures seams or hems with a functional stitch used decoratively. Three embroideries are fragmented pieces from Oseberg (D, K and L) so their finished form and function are not known. The form of stitching has survived on the fragmented piece from Ingleby but its ground fabric has not, so what part of the textile is covered and why is again unknown. This form of functionally decorative embroidery appears sporadically throughout the timeframe, that is, from the 3rd–7th-century Orkney hood to the late 11th–early 12th-century London A top-band.

The data deployed in this section is a useful way of introducing readers to the surviving objects and their constituent parts. It opens up potential areas for further study with regard to work on the significance of correlations between stitch use and design elements. Such information may feed wider debates about changing trends across the timeframe. The data is therefore a good springboard from which to extend the survey's discussion to larger, overarching themes. Nevertheless, the level of information we can derive from the data itself is limited. Moreover, while the pie charts appear to present data in terms of definite statistics, when the information is viewed in tabular format, we see that underlying data does not always bear out such apparent certainties. An example of possible misconception can be seen on Table 1B. The table shows two objects with queries beside them owing to the fact that there is some debate as to the identity of the item on which the embroidery originally appeared. There is no satisfactory way of displaying uncertain data on pie charts, and so the data looks more concrete than it actually is. Such issues have been highlighted where appropriate within the text of this chapter, and they are also covered in particular embroideries' catalogue entries, but the observation underlines the importance of interrogating the data in the pie charts and tables together, and the need for sensitive interpretation. It also emphasises the continuing lack of certainty with regard to particular aspects of the finds. As the discussion at the start of the chapter makes clear, embroidery from this early period survives against the odds. My brief discussion of the incidence of survival of tabby weave as the most common ground fabric for embroidery (pp. 46–47), emphasises the inconclusive nature of some of the data.

From this chapter we can begin to see that the study of embroidery requires us to utilise a variety of approaches, drawing on information from diverse sources and deploying levels of resourcefulness and ingenuity. Plotting correlations between different embroideries and their stitches on charts can be used to begin to generate general theories, for instance, concerned with stitch and material use during the early medieval period and in different geographies within the British Isles. However, dataset size and accompanying uncertainties render interpretation from the figures alone problematic. The next chapter thus follows a different path, seeking to bring to life the unfamiliar subject of early medieval embroidery by conducting a close-up analysis of a little-known individual piece of embroidery, the Kempston fragment, drawing together as much contextual evidence as possible in order to establish a partial object biography of this piece.

Chapter 3

Kempston: the biography of an embroidery

Introduction

This chapter's aim is to begin to assemble an object biography for an individual piece of early medieval embroidery, as a new route to greater understanding of the field. The purpose is to demonstrate that bringing together detailed technical analysis of an individual object, careful study of related attributes and context, and related documentary evidence can shed new light on the bare data. Moreover, as I will go on to show in later chapters, piecing together the life of an embroidery through time helps inform discussions about embroidery's wider social and early medieval contexts.

The Kempston embroidery as it is seen today differs not only from how it functioned when it was originally created, used, and later, buried, but also from how it was perceived when it was discovered in 1864 by the Reverend S. Edward Fitch, and then, nearly 100 years later in the 1970s, analysed by Elisabeth Crowfoot. Due to reasons beyond Crowfoot's control, her results were not made public until 1987, but the research itself is now forty years old and the advent of digital cameras and microscopes facilitates more detailed examination.[1] In particular, my own research shows that the design of the embroidery is different from that postulated by Crowfoot and this affects how we understand the embroidery. The first part of the chapter considers the embroidery today, unpicks scholarly work and changes to the artefact's situation since its discovery, and assesses what we can learn about its medieval context from close analysis and other finds. The second part assembles a provisional biography of

---

[1] Her work was originally presented in a paper at the third North European Symposium for Archaeological Textiles in May 1987, subsequently published as Crowfoot, 'Textile Fragments from "Relic-Boxes" in Anglo-Saxon Graves', in 1990.

the embroidery using the information we can glean from the sources, and discusses the broader implications of such work.

The museum context and what we see today

The Kempston embroidery is made up of seven small fragments of wool embroidery that measure a total of 24 × 54 mm. It is laid out on a piece of plywood board that has been covered with a layer of Melinex (polyester film), with a cushion of polyester wadding on top, encased in conservation fabric (probably cotton). The embroidery has been laid on the cotton and allowed to sink into the polyester cushion to relieve any strain or stress in the fibres. A piece of conservation plastic (probably Perspex), is positioned over the top of the embroidery, held in place with two Perspex clips.[2] The artefact is kept in a pull-out museum drawer system in the store rooms at the British Museum. When a researcher's request to analyse the piece is granted, it is taken out of storage to the study room for the Britain, Europe and Prehistory Collections, for the period of study. The embroidery has also been photographed by the museum using a high-resolution digital camera and the image is available to view via the British Museum's online catalogue.[3]

The embroidery was mounted at some point after it had been studied by Elisabeth Crowfoot in the 1970s and prior to my own visit in the early 2000s. Although the museum has no record of the conservation, the incumbent textile conservator suggests it probably took place before 1994, when she came to work at the British Museum, and the type of clips used to hold objects in place changed.[4] The approach to storage and conservation we see here is good for stabilising embroidery but means it is difficult for researchers to analyse the embroidery in detail as a textile, particularly the reverse, where much of the constructional information is hidden.[5] When Crowfoot analysed the embroidery, we know from her records that

---

2 Pers. comm. Sue Brunning, Curator: Insular Early Medieval and Sutton Hoo Collections, the British Museum (2016).
3 The British Museum, 'Collections Online', *Trustees of the British Museum* (2017) http://www.britishmuseum.org/research/collection_online/collection_object_details.aspx?objectId=86220&partId=1&searchText=Kempston&page=2 (Accessed: 3 July 2018).
4 Pers. comm. Brunning (2016).
5 The Bayeux Tapestry provides a good example of how the reverse of an embroidery provides technical information that helps us understand how an embroidery was created. Photographs of the reverse were analysed, leading to new insight into how the hanging was stitched, the management of threads, and the organisation and skill of the workforce. See A. Lester-Makin, 'The Front Tells the Story, the Back Tells the History: a technical discussion of the embroidering of the Bayeux Tapestry', in *Making Sense of the Bayeux Tapestry: readings and reworkings*, eds A.C. Henderson with G.R. Owen-Crocker (Manchester: Manchester University Press, 2016), 23–40; A. Lester-Makin, 'Les six châteaux de la Tapisserie de Bayeux: Une discussion technique du travail de broderie de la Tapisserie de Bayeux', in *L'Invention de la Tapisserie de Bayeux: naissance, composition et style d'un chef-d'œuvre médiéval*, eds S. Lemagnen, S.A. Brown and G. Owen-Crocker (Rouen: Point de Vues and Musée de la Tapisserie de Bayeux, 2018), 73–91.

it had already been separated from a small copper-alloy container in which it had apparently been found, but there is no evidence as to how it was being stored.[6] Comparison of a photograph taken by Crowfoot and the British Museum's more recent image indicates that re-positioning of the pieces must have taken place when the embroidery was conserved prior to 1994.

The embroidery, and presumably its copper alloy container, came to the British Museum in 1891. In March of that year, a Miss Anne Scott wrote to the museum asking if it would consider purchasing artefacts she owned that were discovered at Kempston, for the sum of £100. In response to a letter she received from Augustus Wollaston Franks (then Keeper of British and Medieval Antiquities and Ethnography), she replied to say she would wait until 11 April 1891, when Franks must have said the matter would be put before the museum board.[7] The board presumably agreed to the purchase because it now possesses the objects, including the embroidery and the container (which is held in a separate part of the collection at the British Museum). Scott may have been a relation of Edward Fitch, who ran the excavation site at Kempston, and perhaps allowed the embroidery and other artefacts discovered during the dig to be dispersed within the Fitch family.[8] It is unclear what happened to the material from the excavated site directly after the excavation, but we are told in an article attributed to D.H. Kennett, that it became disorganised between its discovery and its purchase by the British Museum.[9]

Discovery context

The embroidery was apparently discovered on 18 January 1864, when Fitch noted in his diary that a grave of a woman had been opened at the excavation site of Kempston in Bedfordshire.[10] Crowfoot would later record this as 'probably Grave 71' referring to the conclusion of D.J. Kennett, who, in 1973, published a gazetteer of finds from a number of cemeteries, including Kempston, in which he states that the box and its contents came from this grave.[11]

---

6 E. Crowfoot, 'Textile Fragments from "Relic-Boxes" in Anglo-Saxon Graves', in *Textiles in Northern Archaeology: NESAT III: textile symposium in York, 6–9 May 1987*, eds P. Walton and J.-P. Wild (London: Archetype Publications, 1990), 47–56 (47).
7 Pers. comm. Brunning (2016).
8 We do not have evidence to prove this.
9 D.H. Kennett, 'Recent work on the Anglo-Saxon Cemetery found at Kempston', *South Midlands Archaeology* 16 (1986), 3–14 (4).
10 [Rev.] S. Fitch, 'Discovery of Saxon Remains at Kempston', in *Reports and Papers read at The Meetings of the Architectural Societies of the County of York, Diocese of Lincoln, Archdeaconry of Northampton, County of Bedford, Diocese of Worcester and County of Leicester* (Lincoln: Brookes and Vibert, 1863–1864), 269–299 (291).
11 Crowfoot, 'Textile Fragments', 47, 49; D.H. Bennett, 'Seventh Century Cemeteries from the Ouse Valley', *Bedfordshire Archaeological Journal*, 8 (1973), 99–108 (100). The surname 'Bennett' is a printing error. My thanks to Catherine Hills for drawing my attention to this.

The site of the 1863–1865 excavation was a gravel pit located approximately two miles west of the village of Kempston, Bedfordshire. As the workers were digging out gravel, they found human remains and called in the antiquarian Fitch, who had asked them to inform him of any discoveries they made. The site turned out to be a cemetery containing the graves of men, women and children who had been interred with many personal possessions, including dress accessories, jewellery and weapons, in an irregular configuration of burials. Fitch identified the cemetery as pre-Christian, owing to the different orientations of the buried bodies and the inclusion of grave goods in the graves, and dated the site to between AD 597, when the Augustine mission arrived in England to convert the population to Christianity, and 681, the date when 'the latest pagans of Sussex were secured in the folds of the gospel net'.[12] Kennett's re-examination of the artefacts from the 1960s onwards demonstrated that the cemetery could be dated more precisely to between the 5th and 7th centuries.[13] Crowfoot, who asserts that the grave from which the embroidered fragment was recovered was thought to date to the mid- to late 7th century may be working from Kennett's published gazetteer, which places this grave within a broad 7th-century timeframe.[14]

In his diary Fitch described how a bronze box, which measured 2.75 inches high (68 mm) including the lid, and 2.25 inches in diameter (55 mm),[15] had been discovered by the right leg of the deceased. The outside of the box had remnants of gilding with a punched pattern visible on the lid (Pl. 12d). Near the box was a piece of decayed soft leather or skin, which Fitch thought may have originally been a pouch in which the box and its contents had been kept: a similar box, which had been discovered the previous November in what is assumed to have been grave 64, had a similar product associated with it. When he opened the box from grave 71 Fitch wrote that it contained pieces of fabric, including a fragment of woollen material and scraps of flax linen, which were noted as being 'of three differing qualities'.[16] Although Fitch does not mention the embroidery specifically, Crowfoot wrote that Kennett believed that most of the fabrics found at the site came from this container and that the embroidery was the worsted fabric Fitch had mentioned in his diary.[17] Kennett only states that 'linen and worsted thread' came from this grave in his gazetteer.[18] While the documentary evidence pointing towards the embroidery's discovery in grave

---

12 Fitch, 'Discovery of Saxon Remains', 283.
13 Kennett, 'Recent work', 5–9.
14 Crowfoot, 'Textile Fragments', 47; Bennett, 'Seventh Century Cemeteries', 99.
15 The body of the box measures 2 inches (50 mm) and the lid measures 0.75 inches (119 mm).
16 Fitch, 'Discovery of Saxon Remains', 289, 291–292; Crowfoot, 'Textile Fragments', 49.
17 Crowfoot, 'Textile Fragments', 49.
18 Bennett, 'Seventh Century Cemeteries', 99.

71 is less than conclusive, Kennett's statements and the brief descriptions in Fitch's diary, make it plausible.[19]

The function of the copper alloy box is obscure. A number of comparable boxes have been found in Derbyshire, North Humberside, Yorkshire, Kent and Wiltshire.[20] They have been called work-boxes, relic boxes, and herbal or first-aid boxes because they often contain small samples of organic material such as flax, wool and linen fabrics; possible moss fibres, and silk and woollen threads; decorative edgings such as tablet-woven bands and cords; hemmed fabrics; wooden and metal splinters; vegetable remains; and strung and loose seeds.[21] All bar one have been found in female graves; the exception was located in a child's grave at Updown, Kent.[22] All those found *in situ*, except for the box from Bulford, Wiltshire, were lying next to the left or right hip, suggesting that they originally hung from the waist of the deceased, with other objects usually associated with the chatelaine (a decorative belt that hung from the waist that had useful objects such as keys and scissors attached to it).[23] The Bulford example was discovered placed on the deceased's chest.[24] The boxes are cylindrical in shape, made of copper alloy, and small, measuring on average no more than 120 mm in length and 60 mm across. Each has a tight fitting lid, either hinged or separate, and most have a metal hoop that would enable them to be hung from the belt.[25] Recent research by a group of scholars in Cardiff has been used to create a chronological framework for Anglo-Saxon graves and grave goods from the 6th to 7th centuries, and they have been able to situate the copper alloy boxes within this timeframe, dating them to the 7th century.[26] Examples of the same style have been found in Denmark. These also contain threads and small

---

19 It is even perhaps possible that Crowfoot and Bennett, who were working on overlapping topics at the same period, had discussed their views with each other and so, on the basis of such conversations, Crowfoot felt comfortable ascribing firmer views to Bennett than he adopted in his own writing, but this is only surmise.
20 M. Foreman, 'Work-Boxes (Graves I, II, 183)', in *The Anglo-Saxon Cemetery at Castledyke South, Barton-on-Humber*, eds G. Drinkall and M. Foreman (Sheffield: Sheffield Academic Press, 1998), 285; Anon, '150 Anglo-Saxon graves found at Bulford', *Current Archaeology*, 315 (2016), 6–7; Anon, 'More Saxon graves on Salisbury Plain', *Current Archaeology*, 316 (2016), 8–9. Also see Anon, 'Britain in Archaeology', *British Archaeology*, 149 (2016), 10.
21 Crowfoot, 'Textile Fragments'; C. Fell, *Women in Anglo-Saxon England* (London: British Museum Press, 1984), 40; H. Leyser, *Medieval Women: a social history of women in England 450-1500* (London: Weidenfeld & Nicolson, 1995; repr. Phoenix, 1997), 15.
22 Crowfoot, 'Textile Fragments', 51.
23 S. Chadwick Hawkes, 'The Archaeology of Conversion: Cemeteries', in *The Anglo-Saxons*, ed. J. Campbell (London: Phaidon, 1982; repr. London: Penguin, 1991), 48–49 (49).
24 Anon, '150 Anglo-Saxon graves' (2016), 6–7.
25 A.L. Meaney, *Anglo-Saxon Amulets and Curing Stones*, British Archaeological Reports, British Series, 96 (Oxford: Archaeopress, 1981), 181.
26 A. Bayless, J. Hines and K. Høilund Nielsen, 'Interpretative Chronologies for the Female Graves', in *Anglo-Saxon Graves and Grave Goods of the 6th and 7th Centuries AD: A Chronological Framework*, eds J. Hines and A. Bayless, The Society for Medieval Archaeology Monograph 33 (London: The Society for Medieval Archaeology, 2013), 339–458 (370).

scraps of fabric but none have been found with embroidery.[27] Similar examples from other parts of Scandinavia and Frankish burials are spherical in shape, made up of two sides hinged at the top and latched at the bottom. Some of these also have a metal hoop attached to the top like the English examples. These containers are often richly decorated but, unlike the English ones, they have not been found to contain threads.[28]

Sonia Chadwick Hawkes argued that because the boxes were always found in conversion-period cemeteries, and a number of them, including one found at Updown, Kent, were decorated with a cross, they must be Christian relic boxes containing small relics precious to the owner.[29] Crowfoot came to the same conclusion but qualified her judgement by adding that although during the 6th and 7th centuries Christianity became the dominant religion, certain pagan beliefs may still have persisted.[30] Audrey Meaney suggested that the boxes were meant to be symbolic of a woman's role in society, and as such, took on a role similar to a pagan amulet, or charm. This seems to suggest a dual role as a form of protection and also a confirmation of identity.[31] If we understand elements of Christian and pre-Christian meaning and function to have existed concurrently, the boxes may be a material manifestation of an example of a pagan belief system morphing into a Christian one; the relics of the new faith, contained in the boxes, acting like amulets of the old. Seen in such a light it is not difficult to understand why the boxes and their contents would have been worn not only in life, but buried with the deceased as a form of insurance policy.

What is clear is that the containers cannot be functional work-boxes, although this term has persisted in the published literature, because they are too small to hold all the items used in textile work. Moreover, the fragments often come from pre-existing constructed items that appear to have lost their original use. Boxes which contained vegetable matter or herbs have sometimes been called 'medicinal boxes' because of the possible medicinal properties the organic remains may have had.[32] Perhaps more appropriately, some researchers have started to use the term 'amulet box' which covers a wider variety of symbolic, religious and social options.[33]

---

27 Pers. comm. Ann Zanette Tsigaridas Glørstad, Associate Professor in Archaeology, University of Oslo (2018). For discussions on workboxes discovered in Denmark see C.J. Becker, 'Zwei Frauengräber des 7. Jahrhunderts aus Nørre Sandegaard, Bornholm', *Acta Archaeologica*, 24 (1953), 127–155; C.J. Becker, *Nørre Sandegaard. Arkæologiske undersøkelser på Bornholm 1948-1952*, Historisk-filosofiske Skrifter 13 (Copenhagen: The Royal Academy of Sciences and Letters, 1990); M.M. Hald, P. Henriksen, L. Jørgensen and I. Skals, 'Danmarks ældste løg', in *Nationalmuseets Arbejdsmark* 2015, eds M.M. Hald, P. Henriksen, L. Jørgensen and I. Skals (Copenhagen: National Museum of Denmark, 2015), 104–115.
28 Meaney, *Anglo-Saxon Amulets*, 181, 183.
29 Chadwick Hawkes, 'The Archaeology of Conversion', 49.
30 Crowfoot, 'Textile Fragments', 51.
31 Meaney, *Anglo-Saxon Amulets*, 188.
32 Meaney, *Anglo-Saxon Amulets*, 184–189 (esp. 188).
33 Pers. comm. Gale Owen-Crocker (2011).

Production

An important step in piecing together the life of the embroidery is close analysis of its production: what it was made from and how. Information regarding the materials from which the embroidery is made provide clues about the likelihood that they were locally produced or imported, and considered cheap or expensive, which in turn have implications for the commissioning of the embroidery and how the finished embroidery might have been regarded. A technical analysis of the stitches tells us about the order in which the embroidery was sewn and the level of training its worker had received.

The British Museum's high resolution digital photograph shows that the woollen fragments consist of three pieces (see Pl. 2b). However, in the photograph taken at the time of Crowfoot's research, they are more numerous, with the central section split in two (Fig. 28) and it is not known whether the two central pieces should be joined together or were originally attached via a now disintegrated section.

The remains of embroidery cover the majority of fragments. The three smallest pieces, one on the left and two on the right of the larger central portion(s), are completely covered in embroidery, possibly red in colour. The larger section(s) show clearly defined areas of embroidered and unadorned ground fabric. Measurements of

Figure 28. Line drawing of the Kempston embroidery based on a photograph taken by E. Crowfoot, © Alexandra Lester-Makin

the surviving pieces give approximately 54 mm × 24 mm for the ground fabric with the embroidery as 33 mm wide by from 9 to 1 mm high with individual stitches being 1 mm long. Crowfoot described the ground fabric as a wool twill of broken diamond weave with embroidery threads made of fine plied wool. According to Crowfoot, K. Starkie from the Dyestuff Division of Imperial Chemical Industries Ltd., and, in 1985, G.W. Taylor, a dye consultant, established that the colour of the ground fabric was purple, while the embroidery threads ranged from white (or yellow) and blue (or green) to red.[34] The British Museum photograph, together with my own microscopic images taken during the research trip, indicates that the colours described are as accurate as it is possible to gauge without further scientific analysis.

The microscopic photographs help to establish the order in which the different coloured bands were completed. The order of work can be established by looking closely at the threads and stitching. If threads slightly overlap or impinge on others it is a good indication that those that lay over the top were sewn after those that lie underneath. For instance, when stitches have been caught by other stitches, and the needle has split the thread of one stitch when creating another, the stitch that was caught was clearly embroidered first. The evidence indicates that overall the fragment was worked from left to right, as it is now orientated. In each differently coloured band of embroidery, the stem stitch outlines were worked first, the central filling stitches second.

A technical analysis of the stitch work of the Kempston embroidery reveals a number of interesting facts. The embroidery is very finely worked. Crowfoot judged that it was likely to have been worked in stem stitch, each area being filled in with a double row of stem stitch and outlined in a single row.[35] None of the filling stitches shows clearly, even in the microscopic photographs. However, looking at the embroidery in high resolution and microscopic detail, it appears that at least one of the central areas is filled with a single line of chain stitch, as opposed to stem stitch (Pl. 13a).

At one point, a filling stitch on the over-lying band at the top of the design may tentatively be identified as stem stitch. The microscopic image (Pl. 13b) appears to show a line of stem stitch in the white/yellow thread running from bottom left to top right. Variations in stitch use in different areas may suggest the worker modified her approach to suit circumstance, filling wide central bands with the broader chain stitch and narrower bands with the more slender stem stitch. In turn, this appears to demonstrate the wisdom of working the outlines first and completing the infill afterwards, allowing the type of embroidery stitch chosen to be dictated by the amount of space left in the central area. Although a view of the reverse would be

---

34 Crowfoot, 'Textile Fragments', 52. I have not been able to discover records of more recent tests.
35 Crowfoot, 'Textile Fragments', 49.

needed to confirm this, it suggests a sophisticated utilisation of stitch types on the worker's part.

Generally, the work is very consistent, however, there are two areas of ambiguity – potentially errors. In one place, the microscopic image seems to show a 'split' stitch worked on the white/yellow band lying over a blue/green one (see Pl. 13a). The stitch is located in the curve of the stripe, and is worked into the top of the previous chain stitch. It may be that the thread is thinner at this particular point, giving it the appearance of split stitch when it is actually chain stitch; however, even if we take this into account, the thread at this point does not appear to be plied as the rest obviously is. This suggests that the plied thread has been accidently split by the next stitch, creating a split stitch instead of a chain stitch. Following the split stitch, the band of chain stitch continues until it reaches a break in the design, where the fabric and embroidery have disintegrated. However, the stripe appears to continue along the second fragment which, since it has been placed next to the first during conservation, forms a curve that eventually bends down to the right (Pl. 13c). The embroidered stem stitch outline of this band continues down the curve until it meets the blue/green band running horizontally. The filling here is discontinuous: there is a gap before the downward curve, although a couple of small individual stitches have been worked into the ground fabric in the gap (Pl. 13d). This gap may or may not be another mistake. The embroidery-worker may have missed a couple of stitches here, or pulled the thread too tight during her work, producing two or three very small individual stitches. A third possibility, however, suggested by the complete final stitch before the gap, is that a number of stitches have disintegrated, leaving small casting on/off stitches visible. This would indicate that the embroidery thread ran out at this point and a new one was started. If so, the last complete stitch to the left was the final one made with the old thread before casting off, so it has not unravelled, while the stitches worked after the break are not as complete.

The individual stitches in the Kempston embroidery are fine, measuring approximately 1 mm long, which, as the survey data highlights, is the average stitch length for extant embroidery from this period. Such a fine, consistent technique demonstrates dedication and care in the production process, even in view of the occasional mistakes discussed above. It is not easy to keep stitches so small and neat throughout a piece, and make the design flow at the same time. Indeed, where such small stitches are used, there is a danger that rather than the individual stitches fading into the background and the design and colours coming to the fore, the embroidery looks stilted and disjointed. This is not the case with the surviving Kempston embroidery; the viewer sees the design first and then observes the stitches. The embroiderer must have been trained from an early age, making working an embroidery second nature to her.

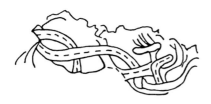

Figure 29. Line drawing of the Kempston embroidery as envisaged by E. Crowfoot, © Alexandra Lester-Makin, after Crowfoot (1990)

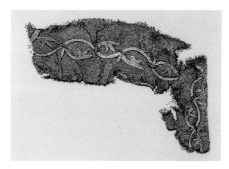

Figure 30. Embroidery from Mammen (measurements not available), © Museum of Cultural History, University of Oslo, Norway

The design

The methods employed on the embroidery – the different stitches apparently used in the wider and narrower bands of infill – indicate that the design was probably drawn on to the ground fabric. This is also evidenced by the accuracy of the completed work. The entwining motif is complex and in order to keep the individual elements in scale and exact, it is likely they were drawn onto the ground fabric. This in turn leads to the inference that the pattern was drawn or sketched out before it was transferred. Pre-sketching the design would allow its intricacy to be honed before it was drafted onto the fabric where, if it was drawn incorrectly, only the embroidery stitches could be used to mask this, leading to the finished piece being skewed and uneven where the worker tried to hide the inaccuracies of the design.

Although the surviving section of embroidery is small, it is possible to discern parts of the extant motif and extrapolate what the whole design may have looked like. Crowfoot suggested that the embroidery formed a scroll border with possible leaves and a bud. She pointed out that although there were no contemporary examples of similar designs in wool embroidery, an example of an acanthus leaf scroll had been found on a cloak from a rich 10th-century grave in Mammen, Denmark.[36] Figs 29 and 30 show Crowfoot's drawing of the Kempston embroidery next to one of the Mammen exemplar. Although this theory has been accepted by researchers since its publication, it raises a number of problems. Firstly, the drawing that Crowfoot made is not completely accurate when compared to the recent high resolution and microscopic photographs; secondly, there are important differences between the Kempston and Mammen embroideries. Finally, the dates of the two embroideries are incompatible: if the 7th-century date for the Kempston fragment is accurate, they are three centuries apart. While it is true that embroidery styles and techniques were often used continuously over long periods of time – indeed, many of the stitches used throughout the early medieval period are still worked today – when combined with the other evidence, the discrepancy between the

---

36 Crowfoot, 'Textile Fragments', 49.

dates of the two embroideries makes Crowfoot's thesis appear more tenuous, creating in this author's eyes a need for further detailed study of the fragment's design.

The first step was to re-draw the embroidery (Pl. 14a). To achieve an accurate representation, the design was traced from a colour print of the British Museum photograph while an enlarged digital version of this and the microscopic images were used for cross-referencing purposes. The result is interesting. The first difference from Crowfoot's drawing is that where she drew what can be interpreted as a single line of stitching, there is in fact a central 'band' outlined on either side with red thread, as she describes in the text. Secondly, where Crowfoot's drawing continues unbroken from the left side through the centre to the right of the embroidery, there is actually a break in the ground fabric, which creates two separate pieces. Despite these being joined together in the photograph, the break is still visible as a discontinuity in the embroidery threads. Thirdly, the original drawing does not show the level of detail that can now be seen in the high resolution and microscopic photographs. The redrawn design is more intricate than Crowfoot's original. As a result, it looks even less like an acanthus scroll.

The coloured threads seem to be used systematically. The red yarn is consistently worked as the outline. The white/yellow thread is used to fill in the band that runs from the left side through the centre, and at the top right-hand side of the design. The blue/green infill appears to the right of the embroidery, lying adjacent to both white/yellow sections. It also appears on the band to the left of the design. It lies partially underneath the white/yellow stripe on the left, and it lies over the white/yellow stripe to the right (Pl. 14b). Combined with more detailed analysis of the order of work, this evidence demonstrates that the blue/green thread is not infilling leaves and buds, and the white/yellow yarn does not define the curve or give definition, as Crowfoot was suggesting. Instead each colour follows its own 'trail' within the design, with different coloured tracks flowing over and under each other at different points. From this reassessment of the evidence, I suggest that the fragment is actually part of an entwined knot or stylised beast motif, similar to those found in metalwork and manuscript illuminations of the 7th

Figure 31. A bracteate die from Castledyke South, Barton-on-Humber, Lincolnshire, © North Lincolnshire Museums Service

Figure 32. Line drawing of the design from the shield strip discovered in mound 1, Sutton Hoo, Suffolk, © Alexandra Lester-Makin

century. The two different coloured knots or beasts wrap under and over each other in a manner almost identical to the entwined knots on the bracteate 'die' (a cast from which metal buttons were made) from Castledyke South, Burton-on-Humber and the shield strip from Mound 1, the Ship Burial at Sutton Hoo, Suffolk (Figs 31 and 32). Significantly, both date to the early 7th century. The design also resembles that of the early 7th-century gold buckle from Mound 1 (Pl. 15a), and the animals that inhabit the borders of a carpet page from *The Book of Durrow*, c. AD 680. The entwined motif is reinforced within the white/yellow section top right. Here, the design forms one 'tail' lying underneath a second tail before terminating. This is also visible on the bracteate die, the Sutton Hoo shield strip, and a pair of clasps from the princely burial mound at Taplow, Buckinghamshire, which again dates to the 7th century (Pl. 15b).

Interestingly, the creatures on *The Book of Durrow* (Pl. 16) carpet page are painted in three different colours, with two for each border: red and yellow across the top and bottom, and yellow and green on either side; a light creature emphasised against the darker one in each case. The same can be seen on the embroidery, where the lighter white/yellow creature entwines itself around the darker blue/green one. The creatures in all the metalwork and manuscript examples have also been given an outline: the manuscript uses a different colour, while an indented line is used to show the demarcation in all the metalwork examples. All the creatures in *The Book of Durrow* are outlined in yellow/white. On the embroidery all the bands are outlined in red. Little of the embroidered design survives, so it is impossible to say for certain whether the motifs are entwined knots or beasts, however, one of the bands seems to be bending around a green tail/band in a similar manner to the 'tail ends' on the bracteate die and Sutton Hoo shield strip (see Figs 31 and 32). Overall, the stylistic parallels help to strengthen the case for dating the embroidery to the 7th century.

Entwined biting creatures and knot work are part of an artistic type formerly known as Salin Style II, and now usually referred to simply as Style II.[37] It is a type that developed out of the earlier Style I which was popular in what is now northern

---

[37] G. Speake, *Anglo-Saxon Animal Art and its Germanic Background* (Oxford: Oxford University Press, 1980), pp. 1–3.

Europe and England early in the early medieval period. Style II developed in Scandinavia during the mid- to late 6th century. As the form spread south into what is now southern Germany and northern Italy, it incorporated aspects of Roman and Byzantine design. It is this multi-cultural version of Style II, with its stylised animals and balanced interlacing, that eventually arrived in Anglo-Saxon England in the 7th century. Here the style took on its own uniquely Anglo-Saxon identity, especially after the post-Conversion period when England was reinforcing its links with Christian Rome and drawing on classical art forms.[38] The examples discussed demonstrate that it was incorporated in metalwork and manuscript illumination. The evidence from Kempston suggests this popular design idea was also utilised in embroidery. Such expressions of a particular fashion or motif are also identifiable in later medieval embroideries where, for example, Gothic architecture is used to frame figures, so it is reasonable to deduce that the exchange of design ideas (such as the entwined knot) between different crafts was taking place earlier in the period as well.

The embroidery evidence sheds intriguing light on a section of the writings of St Boniface (d. 754), the Anglo-Saxon missionary who travelled to Frisia and other Germanic territories to convert the pagans.[39] In a letter he wrote in the 740s to Cuthbert, Archbishop of Canterbury, Boniface, complained of 'the odious superstition of dress' in monks' clothing (*odibilem vestimentorum superstitionem*), before going on to specify 'those dress ornaments [...] embroidered with the widest of borders, decorated with images of worms' (*Illa ornamenta vestium* [...] *latissimis clavis vermium marginibus clavata*).[40] The description of dress ornaments calls to mind the embroidered detachable bands that are mentioned in the *Liber Eliensis*: King Edgar detaches and gives to the monastery similar bands from his boots.[41] Moreover, Christopher Monk has pointed out that the Anglo-Saxons used the vernacular *wyrms* to mean 'serpent, snake and dragon' as well as 'worm'.[42] It is therefore plausible for Boniface to be complaining about knot-like beasts motifs that had been stitched onto bands decorating the monks clothing. The archaeological evidence of the small Kempston fragment seems to provide us with a further test of this translation, since it too is embroidered with a knot/beast motif. It is not known how large the entire piece was originally, but if we take the small size of the surviving motifs into account, it is conceivable that they could have been

---

38 L. Webster, *Anglo-Saxon Art* (London: British Museum Press, 2012), 61.
39 E. Dümmler, ed., 'S. Bonifatii et Lulli epistolae', *Monumenta Germaniae Historica, Epistolae 3*, Merovingici et Karolini Aevi, I (Berlin: Weidmann, 1892), 215–433 (355, lines 18, 19). I am grateful to Gale Owen-Crocker for directing me to this and to Christopher Monk for his guidance on translating the lines in question.
40 *Monumenta Germaniae* gives 'imaginibus' as a variant for 'marginibus', meaning 'with images' or 'with likenesses, further supporting the reading 'decorated with images'. Pers. comm. Christopher Monk (2018).
41 E.O. Blake, ed., *Liber Eliensis* (London: Royal Historical Society, 1962), 293; J. Fairweather, trans., *Liber Eliensis: a history of the Isle of Ely from the seventh century to the twelfth* (Woodbridge: Boydell, 2005), 358.
42 Pers. comm. Monk (2018).

part of a decorative border attached to another garment, perhaps along very similar lines to Boniface's description of dress. Conversely, Boniface's text serves as a form of documentary evidence corroborating that the Kempston fragment's design should be understood as a knot/beast design as opposed to using Crowfoot's earlier hypothesis.

The entwined knot/beast motif was an important element in Germanic mythology.[43] It represented the World Serpent that circled the earth below the sea by biting its own tail. The imagery symbolised, amongst other things, infinity and power, and protection and healing.[44] It is seen in numerous Anglo-Saxon works of art from the 6th and 7th centuries. It also became commonplace in Christian art, where it may have had a purely decorative function, but may also have been appropriated to symbolise the power and protection of the new religion that spread rapidly throughout southern England through the work of the Augustinian mission, a high status example being the Kingston Brooch (Pl. 17a).[45] The Kempston embroidered motif may represent any number of these strands of thinking. It is probable that the motif was thought to bring protection to the person who wore it on clothing and decorative metalwork. Swords were decorated with it even during the conversion and Christian periods: the discovery of entwined beasts on the reverse of a sword hilt from the Staffordshire Hoard, which dates to the 7th century, may demonstrate this (Pl. 17b).[46] In the light of this, we might perhaps wonder if Boniface's complaint about an elaborate decorative design incorporating 'worms' on the monks' vestments involves a level of uneasiness about the appropriation of a pagan symbol within a Christian design.

The biography of an early medieval embroidery

Armed with a range of evidence from the object itself, together with other material culture and documentary sources, we can establish a plausible, if partial, biography for the Kempston embroidery which facilitates greater understanding of both the object itself and its social context. We can bring the evidence to bear on our understanding of particular phases of its life: first, its commissioning, design and creation, second its life as a precious garment, third, its life as a precious

---

43 Although little is known about pre-Christian belief systems in Anglo-Saxon England, scholars believe the early Anglo-Saxon settlers brought a combination of British, Romano-British and Germanic traditions with them. For an introductory discussion on non-Christian belief systems in Anglo-Saxon England see B. Yorke, *The Conversion of Britain 600–800* (Harlow: Pearson Education, 2006), 98–148 (esp. 101–109), and, for an archaeological perspective, N.J. Higham, 'From Tribal Chieftains to Christian Kings', in *The Anglo-Saxon World*, eds N.J. Higham and M.J. Ryan (London: Yale University Press, 2013), 126–178 (esp. 149–153).
44 Speake, *Anglo-Saxon Animal Art*, 90–92.
45 Webster, *Anglo-Saxon Art*, 66–67.
46 Anon, 'Piecing Together the Staffordshire Hoard', *Current Archaeology*, 305 (2015), 6–7.

recycled object and finally, what we know of its after-life as an artefact since its 19th-century discovery.

The materials from which the Kempston embroidery is made suggest a local production. This is a complicated area of study: successive waves of immigrants arrived in England during this period, each bringing their own crafts and technologies, including knowledge involving the production of twill weaves.[47] However, Walton Rogers' research has shown that the particular weave of the worsted wool ground fabric from Kempston – broken diamond twill – is indicative of early Anglo-Saxon production because examples of early diamond twills are always of the broken diamond variety. Moreover, Walton Rogers argues that from the end of the 5th century through to the 10th century, wool twill fabric, particularly in the 2/2 form and using Z twisted/S plied yarn, became the most common type of textile used for clothing in Anglo-Saxon England. The twill pattern appears in its basic form and in various diamond patterns.[48] Owen-Crocker agrees with Walton Rogers but suggests that patterned twills may be rarer and were probably classed as luxuries in contrast to the plain twill weaves. She qualifies this by pointing out that survival rates could well be skewing the data, as examples of patterned twills appear in all the same locations as early Anglo-Saxon burials.[49]

At this period Kempston was situated within the Mercian kingdom near the boundaries with East Anglia and Essex. Mercia was one of the largest and most powerful kingdoms of Anglo-Saxon England.[50] The kingdom's elites had converted to Christianity after the death of the last pagan king, Penda, in 655.[51] Kempston's location in an area of political and economic power bordering other kingdoms suggests the people of Kempston had contact with a variety of travellers, traders and missionaries journeying between regions.

Although we do not know what the embroidery originally decorated, the evidence suggests it was a border worked either directly onto a textile or as a separate band that was later applied to a finished object. What we know of the symbolism of serpent forms suggests that if the piece decorated a garment of some kind, the motif – and indeed the garment – were probably regarded as protective: either owing to the serpent's pagan significance and/or as a motif appropriated to Christian symbolism. The documentary evidence from Boniface allows us to tentatively suggest that the

---

47 P. Walton Rogers, 'Cloth, Clothing and Anglo-Saxon Women', in *A Stitch in Time: Essays in Honour of Lise Bender Jørgensen*, eds S. Bergerbrant and S.H. Fossøy (Gothenburg: Gothenburg University, Department of Historical Studies, 2014), 253–280.
48 P. Walton Rogers, *Cloth and Clothing in Early Anglo-Saxon England: AD 450–700* (York: Council for British Archaeology, 2007), 65; Walton Rogers, 'Cloth, Clothing and Anglo-Saxon Women', 255.
49 G.R. Owen-Crocker, *Dress in Anglo-Saxon England, revised and enlarged* (Woodbridge: Boydell Press, 2004), 293.
50 Higham, 'From Tribal Chieftains', 158.
51 K. Leahy and R. Bland, *The Staffordshire Hoard* (London: British Museum Press, 2009; repr. 2010), 14.

original owner may have been someone with a religious vocation, whether pagan or Christian.

As the century progressed and elite social groups gathered more economic and political power, individuals were able to commission fashionable goods influenced both by religious belief and secular trends. As we have seen, both the Germanic paganism of the ancestors and new symbolism associated with Christianity may coalesce in the design of this piece, demonstrating the morphing of one belief to the other.

We can also deduce that the original owner must have been an important person in the eyes of the individual with whom the fragment of embroidery was eventually buried. Perhaps he or she was a highly-esteemed pagan, and so the recycled embroidery functioned as a personal amulet wielded for the protection of its owner. Or, if the fragmented embroidery is understood as some form of contact relic carried in its copper alloy box, the original owner may even have been venerated as a Christian saint, with the embroidered object functioning not only as protection, but as a resource for augmenting prayer and gaining the ear of the saint (and, in turn, God).

The embroidery is in wool. Wool was the staple fibre of the early medieval period and would have been found across all spheres of society. However, the Kempston embroidery was worked in the finest grade of wool. Moreover, fine worsted wool can create a sheen which may make it look like a much richer material once embroidered. This, combined with the use of purple to dye the ground fabric suggests that the individual who commissioned and perhaps also wore the embroidery (it is also possible that the piece was commissioned by one person as a gift for another) may have stipulated a design that would imitate the more exclusive silk fabrics that were beginning to trickle into the country through ecclesiastical networks. The sheen of such materials would also have meant that as the wearer moved, the light reflecting off the decoration would give the impression that the entwined knots/beasts were moving, imparting an additional aura to the wearer.

Whoever the owner was, the expert stitching demonstrates that the embroidery was a product that would have taken time and expertise to create. The stitching is fine and well executed, demonstrating the technical skill of the worker and her understanding of the materials. She made use of stem stitch, one of the most commonly used stitches in the survey, and chain stitch. Stem stitch's versatility means it produces an outline that does not distract from the finished design, and particular stitches and motifs may be picked out by varying the thickness and colour of the thread. Its flexibility also means it is useful in designs that contain curves and other awkward areas. While it is not a stitch to be used to fill large areas (the coverage achieved through each stitch, or line of stitching, is small), it is useful for filling smaller shapes. Analysis of visible areas of the stitching appears to show that the embroiderer may have modified stitch choice according to the size of the area to be filled, using chain stitch to fill larger areas.

Although this adaptive approach points to a high level of expertise, evidence does not indicate that a structured framework of training, such as apprenticeship, was in place this early in the early medieval era. Girls, and the surviving evidence points solely to women embroiderers during this period, were taught to embroider from an early age. It would have been one of their tasks alongside other textile oriented and domestic work, and it was a skill that would have been passed down from mother to daughter or elder to youngest as part of everyday life. As such, many women would have become adept at decorating textiles and as settlements developed into larger estates, particular individuals may have specialised in this work (for further discussion, see Chapter 5).

According to my hypothesis, then, after serving an initial period as the garment of a high-status individual with a religious vocation, the archaeological evidence – that is, the embroidery's fragmentary status and position within a female burial, the proximity of the container, and the incidence of other small textile fragments – demonstrates that the embroidery was recycled. This seems to be the case with many examples of textiles and embroidery discovered in archaeological contexts. It was a common phenomenon because textiles were one of the most time consuming and, therefore, expensive products to produce. They were therefore used until they were no longer viable in any form. Once a textile was no longer fit for its original purpose, it would be appropriated to a new use, probably by a new owner of lower status and means. The presence of embroidery gave added appeal to the discarded garment or object. It is conceivable that this small, precious fragment kept safe in its copper alloy box is the last in a succession of a series of recyclings, and that on each occasion the Kempston embroidery's meaning and/or identity was subtly modified. The fact that such a small piece was found with the copper alloy box suggests that the embroidery continued to hold a protective value long after it had finished its initial functional use, reinforcing the thesis that the original owner was held in high esteem, whether Christian or pagan, and/or that the meaning of the entwined serpent/knot motif was viewed as sufficiently powerful to protect long after the textile had been downgraded from its original purpose. It was housed in such a way as to enable it to be carried and kept close to its new owner, acting as a constant reminder of its original form and use, and its continued significance, even in death.

The third phase of the Kempston embroidery's biography is its period underground, about which we can make certain deductions as to the conditions of the burial site and how they have acted on the embroidery. As Chapter 2 highlights, the survival of an embroidery depends on how the fibres used in its production interact with the surrounding environment. The Kempston embroidery, which is made from protein-based animal fibres, had excellent conditions in which to survive, being stored in a small, sealed copper alloy box buried in gravel. Due to the tight-fitting lid of the box, no air was able to penetrate. The gravel in which the inhumation was situated

---

52 Although Fitch did make specific notes about individual pieces of textile in his article and diary, which was an unusual practice at the time. See Fitch, 'Discovery of Saxon Remains', 271, 280, 286, 289, 290, 291, 203.

was free-draining and stayed dry, so that the copper alloy box had little contact with moisture and very little corrosion took place, again protecting the embroidery contained within (see previous discussion from pp. 31–32). The fourth phase is the embroidery's discovery in 1864. The paucity of detail regarding the embroidery in the work of Fitch can be seen as a demonstration of the low status accorded to textile and embroidery as cultural signifiers during the 19th century, aspects of which still provide difficulties.[52] The burial site excavation as a whole, however, provides important details as discussed above, allowing us to reconstruct aspects of the textile's significance as an amulet during the deceased person's life, and confirms the strong association between amulet boxes and female burials. Unfortunately, it is not possible at present to divine the status of those buried within the Kempston cemetery, but the objects that were discovered, including weapons and vessels in probable male graves, and beads, broaches and chatelaines within probable female burials,[53] indicate at least the localised social status and hierarchies of individuals from the settlement that the cemetery was associated with. It should also be remembered that objects such as weapons do not mean that the individual was a warrior. It is now thought that the inclusion of such objects could actually have been intended as markers of age, gender, social status and possibly ethnicity. The same inferences can therefore be drawn with regard to objects included in female burials, for example, the chatelaine on which the copper alloy box that held the embroidery would have hung. Although Helen Geake has observed that apart from them being associated with female burials, other theories about social status and meaning cannot be substantiated. She did tentively suggest that chatelaines dating to this period could have had a functional use while those dating to the 6th century appear to have been symbolic.[54] Women were often buried in their finest clothing with a number of their possessions so that onlookers could 'read' who the person had been and what status and role she had played within her community.[55] For example, the inclusion of this copper alloy box, with its particular style and what appear to be associations with Denmark, may have told onlookers that this woman had connections with people in that geographical area (see discussion on Sutton Hoo B, the cuff, for similar themes, from p. 82). There is less to say about the layout and location of the cemetery: Howard Williams has highlighted that early inhumation cemeteries did not follow fixed orientations or formal layouts, and that they probably simply followed the topography of the land.[56]

---

53 Kennett, 'Recent Work', 6–9.
54 Helen Geake, *The Use of Grave Goods in Conversion-Period England, c. 600–c. 850*, British Archaeological Reports, British Series 261 (Oxford: Archaeopress, 1997), 57–58.
55 H. Williams, 'Mortuary Practices in Early Anglo-Saxon England', in *The Oxford Handbook of Anglo-Saxon Archaeology*, eds H. Hamerow, D.A. Hinton and S. Crawford (Oxford: Oxford University Press, 2011), 238–265 (249–251) for an introductory overview on this subject.
56 Williams, 'Mortuary Practices', 249.

The fifth phase of the biography is the period between discovery and Anne Scott's donation of the embroidery, about which we know little except from Kennett's tantalising comment that the material became disorganised. The evidence from Anne Scott's correspondence with the British Museum highlights the then accepted notion that excavation finds were classed as the private property of the lead antiquarian and other associates linked to a dig, to be disposed of as they saw fit, unless an object was made from precious metals, in which case it was classified as treasure, in a similar manner to modern-day stipulations, which are enshrined in the Treasure Act.[57] This would account for the fact that the British Museum apparently had to buy the Kempston objects from Anne Scott. Finds and ownership are more tightly controlled under modern legislation. Finds are usually the property of the land owner unless they come under the above-mentioned Treasure Act, which now encompasses a wider range of find types, as well as finds from Scheduled Ancient Monuments. Landowners are normally invited to give excavation finds to an appropriate local museum for analysis and curatorship – although they do not have to agree to this and may keep or sell such objects.[58]

The sixth phase is the museum context. This gives us insight into changing priorities around the conservation of textiles. Although the surviving archival evidence is limited, the correspondence between the British Museum and Elisabeth Crowfoot demonstrates that she was regarded in much the same way as her mother, Grace, had been: as part of a network of experts who knew each other and could be trusted with fragile objects. Moreover, rules on handling such objects were far more lenient: for example, she was allowed to unpick fabrics in order to work out their patterns and in some instances she was permitted to snip and keep threads from textiles in order to compare them to other extant examples, a practice that would not be allowed today. She thus had considerably more freedom to study textiles in a hands-on way than would be possible for modern scholars working outside conservation departments.

We might understand the point at which my own research on the textile began, utilising up-to-date technology and attempting to understand it as part of a wide-ranging survey of early medieval embroidery, as well as piecing together as much as possible of the object's timeline, as the beginning of a seventh phase in the embroidery's biography.

Today the embroidery is kept in optimum conditions for its conservation, although this has its drawbacks. While Crowfoot was able to examine both front and reverse of the embroidery when handling it, today such an approach is not possible because the fragment is secured to its protective ground fabric by conservation plastic which

---

57 Last revised 2010. Portable Antiquities Scheme, *Summary Definition of Treasury* https://finds.org.uk/treasure/advice/summary (Accessed: 13 August 2018).
58 Pers. comm. Melanie Giles (2018); Portable Antiquities Scheme, *Summary Definition of Treasury* https://finds.org.uk/treasure/advice/summary (Accessed: 13 August 2018).

is clipped in place. The plastic cover results in the minimisation of any sensory engagement between viewer and embroidery, while measurements, photographs and microscopic images are also restricted by the cover, which, if the researcher is not careful, leads to distortions in the results. Researchers see the embroidery laid out as the conservator has interpreted it, thus 'reading' its arrangement through the eyes of another, as opposed to being in a position to form their own judgements. Moreover, the embroidery has become entirely detached from its original significations, although aspects of how we see it today, such as its fragmentary status and its detachment from the container in which it is believed to have been held prior to its discovery, attest to earlier phases of its biography. Today its cultural significance lies in its status as one of the very few surviving examples of embroidered early medieval material culture from the 7th century. The high technical standard of the Kempston embroidery means it will continue to be of particular interest to researchers. Moreover the valuable information we can glean from this and other fragments of embroidery will continue to help scholars to answer research questions on the early medieval period, and learn more about life 14 centuries ago. The next chapter utilises both the dataset discussed in Chapter 2 and aspects of object biography methodology as deployed in this chapter, to broaden the field of enquiry and establish the significance of embroidery within the development of Anglo-Saxon society.

Chapter 4

Embroidery and Anglo-Saxon society

In the previous chapter of the book I demonstrated the way in which assembling the biography of an embroidery in close-up through detailed study helps us to appreciate how individual material objects can shed light on broader social and cultural developments. In this chapter, I use the technical and biographical data pertaining to each embroidery in the cohort under examination to elaborate on how the embroideries relate to three main phases of development in Anglo-Saxon society, shedding light on cultural values.

An object such as an embroidery is affected by the social and cultural conditions in which it takes shape. Particular physical and cultural circumstances (relative plenty or scarcity of materials, or access to materials, for instance) affect how an object is made, its function, and whether it is re-used by others after its initial purpose has become obsolete, or simply discarded. Re-use with modification, and changing perceptions of an object are also influenced by changing political, social and artistic ideas. We can thus explore how context influences the choice of design and materials we see in the finished embroidery. Conversely, where information about early Anglo-Saxon society is sparse, we can glean important and useful information about the circumstances in which a piece of embroidery was created, using such information to bolster existing evidence, or contradict theoretical surmises.

For the purposes of tracing how social and cultural developments are reflected in early medieval embroidery, we can divide Anglo-Saxon society into three broad phases, however, we need to keep in mind that these phases were not discrete: one overlapped the next and each contributed to the overall development and dissemination of people and ideas forming the melting pot of ethnicity and religion that made up the early medieval society in which these embroideries were conceived and created. The first phase begins with the Anglo-Saxons' arrival in England in the 5th century. These tribal groups brought with them their own cultural ideas and pagan religion. Their influence waned under the second phase, the conversion of the Anglo-Saxons to Christianity in the 7th century, but experienced new impetus with the arrival of the Vikings in the 8th century. Phase two was two-fold, involving on the one hand, the conversion of the northern kingdoms to Irish Christianity and on the other, the adoption of the European Roman form of the faith within the southern kingdoms. Elite and royal families engaged not only with the faith itself but with the trappings of its associated secular world. They allied themselves to Christian elites on the Continent through trading and diplomatic connections and marriage ties, cementing both religious and secular connections with the Roman and Byzantine worlds. Thus, these Anglo-Saxons came to see themselves as the successors to both Rome and Byzantium, an idea they expressed through material culture, including embroidery. The third phase involved the absorption and appropriation of different ideas and influences within Anglo-Saxon society, which became a uniquely Anglo-Saxon expression. This was subsequently disseminated back to areas from whence it originally came, a religious example being the 8th-century missionaries who travelled to Germanic territories to convert the local populace. Each of these three phases involved the evolution not only of cultural and religious ideas, but design and technical advances within artistic and craft circles. This chapter briefly investigates the influence of these overlapping phases in relation to the creation and development of our cohort embroidery.

As with the permeable nature of secular and religious borders that allowed the movement of objects and the potency of new ideas and styles, the same was also true of their influence on Anglo-Saxon embroidery. While there are pieces that display a distinct Scandinavian or pagan (phase 1) heritage, for example, the use of laid-work on the Bayeux Tapestry has links to Scandinavia, or a clear Christian (phase 2) ancestry, such as the metal thread work on silk ground fabrics and the use of religious symbols on the Durham and Worcester embroideries, many examples show the influence of practices associated with more than one phase. A good early example is the Kempston fragment, which is embroidered with a knot or entwined beast motif drawing on the artistic form 'Style II', as Chapter 3 discusses. This style developed in Scandinavia during the mid- to late 6th century. As it spread out from northern Europe through what is now southern Germany and into northern Italy, it assimilated popular ideas and motifs from these areas as well. It was this international version of Style II that arrived in Anglo-Saxon England during the 7th century,[1] where it was used in the design on the Kempston embroidery. This is a prime example of phase 3 absorption, re-configuration and use.

Conversely, some items in the cohort are not easily categorised in this way, for instance, the surviving sections of footwear where it could be argued that the embroidery is generic as opposed to specific to particular phases. On the other hand, it might be argued that the shoe from Coventry, which has an embroidered vamp stripe of raised plait stitch (see Glossary), shows Scandinavian (phase 1) influences, because the use of this stitch seems to be confined to the North Sea Zone.

---

[1] L. Webster, *Anglo-Saxon Art* (London: British Museum Press, 2012), 61.

Phase 1

The number of surviving embroideries influenced by phase 1 traditions is greatest, firstly, for the period when the Anglo-Saxons moved to England and before they converted to Christianity (5th–7th centuries) and secondly, during periods of Germanic and then later Scandinavian contact (8th–10th centuries). Influences are identifiable through the type of stitches, threads and designs incorporated into the embroideries. The earliest such example is the Orkney hood, which has been radio-carbon dated to between *c.* AD 250–615.[2] This textile, originally part of a larger piece, was at some point in its biography recycled and turned into a child's hood. A patch was attached to either side of the hood. Over both of these patches, and protruding beyond them, were stitched double rows of chain stitch in dark brown and yellow wool, which have partially disintegrated. As a finishing touch a tablet-woven band with fringe was stitched to the bottom of the hood using what Audrey Henshall called 'a regular stitch', which was then covered with a complex looped stitch.[3] Both these stitches are, as far as the surviving evidence shows, associated with, first, Germanic and then Scandinavian needlework.

Marjory Findlay, the conservator, classified the double-rows of chain stitch as decorative because they had not been deployed to attach the patches to the hood.[4] Although it is true that both stitches on the Orkney hood are decorative, they also have a functional role, adding extra support at potential weak points. Moreover, their decorative aspect is functional since they hide the seams; a function that can be identified with regard to a number of the surviving embroideries (see below). There is a single example of chain stitch from the 7th century, the Kempston fragment, where it is used solely decoratively. Then the stitch disappears from the archaeological record for the next three centuries appearing again as a purely decorative stitch on the late 10th-century York pouch (York A) where it has been used to create a Latin cross. It is then seen again on the 11th-century Bayeux Tapestry where it has been used on a small section of the hanging worked in both single and double rows of embroidery.

The complex looped stitch used on the Orkney hood exists in a long tradition of stitches used to cover and reinforce seams decoratively.[5] To this can be added other stitches such as the buttonhole used to join the seams of the Utrecht *bursa* together,

---

2 R.E.M. Hedges, R.A. Housley, C. Bronk-Ramsey and G.J. van Klinken, 'Radiocarbon Dates from the Oxford AMS System: Archaeometry datelist 16', *Archaeometry*, 35 (1993), 147–167 (155).
3 A.S. Henshall, 'Early Textiles Found in Scotland', *Proceedings of the Society of Antiquaries of Scotland*, 86 (1951-2), 1–29 (13, 14).
4 M. Findlay, 'Report on the Conservation of the Orkney Hood', in *The Laboratories of the National Museum of Antiquities of Scotland 2*, eds T. Bryce and J. Tate (Edinburgh: National Museum of Antiquities of Scotland, 1984), 95–102 (98).
5 See A. Lester-Makin, 'Looped Stitch: the travels and development of an embroidery stitch', in *The Daily Lives of the Anglo-Saxons*, eds C. Biggam, C. Hough and D. Izdebska, Essays in Anglo-Saxon Studies 8 (Tempe Arizona: Arizona Center for medieval and Renaissance Studies, 2017), 119–136.

a comparative piece from Mammen, Denmark, where raised plait stitch has been used to decoratively cover a seam, and possibly the now lost scrap from Mitchell's Hill where blanket stitch was used to reinforce the edge of the fabric and the five circles on Oseberg C, which, if my hypothesis is right, hung loose over a seam or the edge of the fabric decoratively covering them (see Oseberg C). Looped stitch can be found throughout the early medieval period but appears in the British Isles and Ireland during the early and later centuries only. All extant examples appear in locations associated with, first, Germanic tribes and then Scandinavian trading hubs. I have argued elsewhere that in Anglo-Saxon England, the use of looped stitch died out when elite circles turned from their pagan roots to Christianity, and their focus shifted from the material culture of the north to that of the Christian world. I have also suggested that looped stitch died out because of what the stitch represented.[6] For it may be that looped stitch was thought to have amuletic properties. It is not far-fetched to think of the completed stitch as having a serpent-like appearance; the woven effect of the finished stitch as it encircles the textile it has been worked on looks a little like the overlapping scales of a snake's skin. The same can be imagined with regard to the knot-work and entwined beast motifs found on Anglo-Saxon metalwork and sculpture, and in illuminated manuscripts. If looped stitch was seen in this way, it may also have been believed to take on the qualities of the World Serpent that encircled the earth under the sea by biting its own tail. This Germanic mythological creature represented a complex of ideas including infinity, power, protection and healing. So if looped stitch was viewed as an expression of the World Serpent it may have been thought of as a protecting force, especially at areas of weakness such as the seams and edges of textiles.[7] It was perhaps partially as a result of such an association that looped stitch continued to be utilised in areas of pagan belief and was re-spread to England and Ireland when Viking Scandinavian people re-established links and influence through the Great Army at Ingleby, and trading hubs at York and Dublin. Conversely it may have fallen out of use as Christianity became more influential. See also discussion of the World Serpent in relation to the Kempston fragment (Chapter 3).

There is another stitch that appears to have developed within the Scandinavian world: laid-work (see Glossary). This stitch has been found on two pieces made in Anglo-Saxon England, the Bayeux Tapestry and Durham F, both of which date to the 11th century. The laid-work on the Bayeux Tapestry is constructed of wool stitched onto a linen ground fabric and as such, would have been a quick stitch to work. The efficient way it is constructed, with economical use of thread, makes it good for covering big areas within a design on large textiles made for public display. By contrast, the laid-work on Durham F (see Pl. 9a) was worked in silk floss and thread. This form of laid-work is found on later examples of *Opus Anglianum* embroidery and may therefore show a development from the use of wool to the use of silk. It also shows the development in technique – moving from larger stitches and an open finish

---

6 Lester-Makin, 'Looped Stitch', 135.
7 Lester-Makin, 'Looped Stitch', 135.

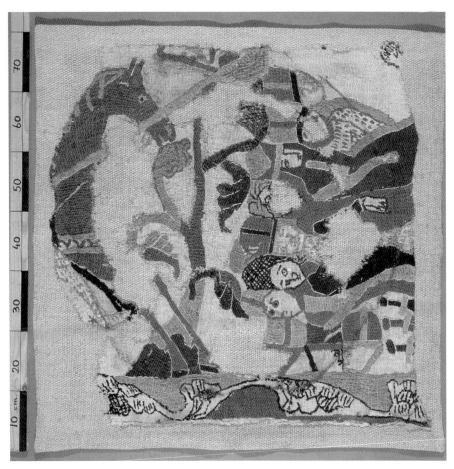

Figure 33. The embroidery from Røn, Norway, © Museum of Cultural History, University of Oslo, Norway

to finer more compact stitch work, which is better suited for smaller elements within a design. The finished stitch would also look more sumptuous when finished. This form of laid-work is a fine, delicate stitch and was used on intricate areas of smaller textiles. It was also more time consuming to produce than laid-work constructed from wool (see also Chapter 5). Durham F can therefore be seen as a development from an early medieval form of laid-work to the later *Opus Anglicanum* technique. All other surviving examples of laid-work worked in wool on a linen ground fabric date to after the early medieval period and all appear on embroideries from Norway and Iceland. The earliest surviving piece from Norway, dated to the 12th century on the basis of the foliage design, is the Røn fragment (Fig. 33).[8] All the published examples from

---

8 G. Wingfield Digby, 'Technique and Production', in *The Bayeux Tapestry: a comprehensive survey*, ed. [Sir] F. Stenton (London: Phaidon Press, 1957), 37–55 (49).

Iceland (also all in wool on linen, though some later pieces incorporate gold thread) are later in date, ranging from between the second half of the 14th century to the second half of the 17th century.[9] It appears that laid-work in wool on a linen ground was part of another long tradition that continued well past the early medieval period.

The relatively sophisticated use of laid-work on the Bayeux Tapestry demonstrates that the stitch was known before the hanging's construction. The Tapestry exhibits no development of the stitch at a constructional level, indicating that the embroiderers were familiar with how to create it. The same is true for the small surviving section on Durham F. The present evidence for laid-work as discussed here, indicates that the wool on linen version of the stitch was of northern origin and spread throughout the North Sea zone but not beyond, possibly in a similar manner to looped stitch. As with looped stitch, it may be that laid-work's use began to expand and develop along different lines, particularly in elite circles in conjunction with the use of more expensive metal threads in embroideries, and it is this we see on Durham F. This variation may well have spread beyond the Scandinavian world when it was incorporated into *Opus Anglicanum*.

It was not just stitches that embodied multi-level meanings. As already discussed in relation to the Kempston fragment, it is probable that the same applied to design, particularly when this helped express affiliation to a group or ancestral home. There are two examples in the cohort: the 6th-century fragment from Mitchell's Hill and the 7th-century embroidery, Sutton Hoo B. Both objects are associated with cuffs.

The Mitchell's Hill stitching was discovered positioned between one of a set of wrist-clasps (see Fig. 8). A set of wrist-clasps was used to hold the cuffs of one sleeve together. Therefore each set of clasps came with a partner set, so that both sleeves of a garment could be secured. At the time of John Hines' publication on the Mitchell's Hill clasps in 1993, he noted that they were a unique surviving example, and this is still the case today.[10] Although the clasps had no dateable context, Hines demonstrated that their design was an amalgamation of motifs found on other clasps from the same typology, type B18. This particular set consisted of a ribbed bar on one of the clasps only. A row of round-based knobs was to be found on the reverse of both, between which were located lugs, small protruding bars through which sewing thread was passed when the clasps were attached to a garment. At the top of one of the clasps was a cast decorative triangle.[11]

Originating in Scandinavia, Hines has discussed in detail the influences and geographical spread of Class B clasps, concluding that the type seen in the east of England came from Norway, probably from as early as the last quarter of the 5th

---

9 See E.E. Guðjónsson, *Traditional Icelandic Embroidery* (Reykjavik: Iceland Review, 1985).
10 J. Hines, *Clasps Hektespenner Agraffen: Anglo-Scandinavian Clasps of Classes A-C of the 3rd to 6th centuries AD. Typology, Diffusion and Function* (Uddevalla: Bohusläningens Boktryckeri AB, 1993), 2, 59, 61–62, 63. He classifies them as type B18 h.
11 Hines, *Clasps Hektespenner Agraffen*, 61–62.

century and definitely during the 6th century.[12] Of note is that while those discovered in Norway are usually associated with male burials, or those assumed to be male, and located in areas of the body other than the wrist, those from Anglian England are almost exclusively from female contexts and used only as wrist-clasps.[13]

Hines suggests clasps formed part of a material culture brought to Anglian England from Norway through migration and as the majority of clasps appear in East Anglia, it would seem probable that they are indication of a communal migration. In England the clasps formed part of a hybrid culture that systematically developed within the area the migrants colonised. Hines goes on to hypothesise that during the 6th century the role of clasps as part of this material culture became consolidated via communal consensus (although perhaps not a conscious consensus) into part of a 'national Anglian dress', reflecting a group identity.[14] It is as part of this morphing and cementing of cultural identity that the Mitchell Hill clasps would have been created, worn and deposited.

While the stitching at Mitchell's Hill is associated with a pair of physical wrist-clasps, Walton Rogers has suggested that Sutton Hoo B is intended to represent and mimic cuffs constructed from tablet-woven bands and metal studs (see Fig. 5).[15] Such a hypothesis is reasonable and can be further developed. The embroidered cuffs are extremely similar to a set of small metal button clasps discovered at Øvstebo, Vindafjord, Ro in Norway (now lost), which are thought to date to c. mid-5th century.[16] Hines' drawing of the button clasps from Øvstebo (Fig. 34) shows two vertical lines of four buttons, one line on each edge of a cuff, forming one set of clasps that held the cuff closed. The Norwegian example was also decorated with small circles in the centre surrounded by two rings that look like the stem stitch embroidery on the Sutton Hoo example. Here, as with the Mitchell's Hill example, the link to Norway supports the hypothesis for the spread of this form of clasp as part of migration into eastern England. In Scandinavia this form of clasp was used primarily, if not exclusively, by men; but like the Mitchell Hill clasps, once they had become part of the Anglian 'ethnic' dress, they were adapted to become part of the female dress code.[17]

Seen in this light the embroidered cuffs and 'buttons' from Sutton Hoo can be understood as a decorative development of elite dress that cemented the wearer's status as a rightful member of the leadership group. The embroidery did so by referring back to the use of actual clasps and buttons, as found in the archaeological record for Anglian dress. It thus affirmed the origins of the group in Scandinavia,

---

12 Hines, *Clasps Hektespenner Agraffen*, 15, 16, 92.
13 Hines, *Clasps Hektespenner Agraffen*, 81.
14 Hines, *Clasps Hektespenner Agraffen*, 92.
15 P. Walton Rogers, 'The Textiles from Mounds 5, 7, 14 and 17', in *Sutton Hoo: an Anglo-Saxon princely burial ground and its context*, ed. M. Carver (London: British Museum Press, 2005), 262–268 (266–267).
16 Hines, *Clasps Hektespenner Agraffen*, 32, 33, contains useful detailed discussion of the button clasps.
17 Hines, *Clasps Hektespenner Agraffen*, 78, 81.

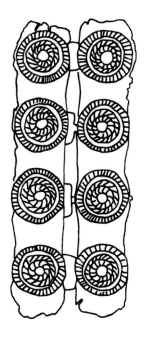

Figure 34. Line drawing of metal button clasps from Øvstebo, Vindafjord, Ro in Norway, © Alexandra Lester-Makin, after Hines (1993)

whilst demonstrating that the wearer had assumed a new Anglian identity along with the social group of which she was a part. According to Hines, the clasp as a marker would have been helpful to a nascent political class looking for new forms to express itself. He argues that as aristocratic lordship became consolidated, the use of clasps as a form of group identity mattered less. Thus, as part of a process of social and political change, clasp use became more controlled, and the volume and type of clasp being produced and circulated diminished. The idea of an Anglian dress was gradually abandoned since the established society no longer felt the need to express its links to its Norse homeland, once it was fully settled and integrated into the Anglian territory.[18]

It is not just stitches and designs that appear to have been favoured by people of Germanic and Scandinavian heritage. Similar trends can be seen in the use of particular materials, such as pulled metal wire. The fine, pure silver pulled wire used to create the looped stitch of the Ingleby fragment is such an example (see Pl. 7b, c). This type of metal thread has been found on six examples of embroidered looped stitch from Birka. Two of these were identified by Agnes Geijer as fine silver wire, one was coarse and the other three were not identified in her published work.[19] Other objects made of pulled metal wire such as a possible appliqué from St Mary's Hill Senior, York, and a possible braid and 18 miniature balls from St Patrick's Isle, Peel on the Isle of Man, have all been found in locations associated with Scandinavian control.[20] Geijer has argued that this type of metal thread was produced by the Sami people of northern Sweden, and it is from here that the only evidence for production has so far been discovered. Geijer points out that the merchants and craftspeople of Birka traded with this tribe,[21] so it is not unlikely that this is how the wire came into the possession of Scandinavians and was made into the elaborate ornaments and embroidery found in

---

18 Hines, *Clasps Hektespenner Agraffen*, 94, 95.
19 Unpublished letter from 22 February 1957; Agnes Geijer, *Birka III: Die Textilfunde aus den Grabern* (Uppsala: Kungl Vitterhets Historie de Antikvitets Akademien, 1938), 111–114.
20 R.A. Hall, 'A Silver Appliqué from St Mary Bishophill Senior, York', *The Yorkshire Archaeological Journal*, 70 (1998), 61–66; J.A. Graham-Campbell, 'Tenth-Century Graves: The Viking-Age Artefacts and their Significance', in *Excavation on St Patrick's Isle Peel, Isle of Man, 1982-88 Prehistoric, Viking, Medieval and Later*, eds A.M. Cubbon, P.J. Davey and M. Gelling (Liverpool: Liverpool University Press, 2002), 83–98 (88).
21 A. Geijer, 'The Textile Finds from Birka', *Acta Archaeologica*, 50 (1979), 209–222 (216).

the archaeological record. This evidence in turn, combined with the fact that no pulled metal embroidery has been discovered outside Viking controlled areas, suggests that the Ingleby embroidery was not produced in Anglo-Saxon England but came from a Scandinavian trading hub with access to pulled metal.

While the movement of pulled silver wire can be traced through Scandinavian trading hubs, there are a number of embroidered fragments from Oseberg that cannot be tracked so easily. Arne Emil Christensen and Margereta Nöckert have suggested that the wool ground fabrics of Oseberg A–D originally came from Ireland.[22] As these embroideries date to the late 8th to 9th centuries, they were most likely made before the trading hub at Dublin was established, in 841. A probable hypothesis, therefore, is that the fabric was taken during a raid. However there is no evidence as to where the fabric was embroidered. As I will go on to discuss in Chapter 5, the weaving and stitching of textiles in Ireland seems to have been well regulated, with strict codes as to who could participate in particular activities. It is therefore possible that the textile fragments at Oseberg were woven in one place and embroidered in another location, before the finished fabric was removed abroad during a raid. However, there are alternatives: the textile may have been seized by Viking voyagers before it was embellished and embroidered when it arrived in Norway; or it is possible that the woven textile found its way to Anglo-Saxon England through trade, was there embroidered and only then taken in a raid. Thus Oseberg A–D not only highlight the difficulties involved in tracking objects, but serve as a reminder that textiles, especially when embroidered, were highly sought after objects.

Phase 2

As rulers of the Anglo-Saxon kingdoms began to convert to Christianity they looked to the Continent for role models for strong leaders within this new world ideal, for example, Francia.[23] However, it was not only the faith of such figures that the elite wanted to mimic but also the secular fashions of conduct and dress. This change was brought about in part by increased contact through trade and imports with new and exotic goods such as silks. Such items were given as gifts to those in elite and royal circles in Anglo-Saxon England. For those who were unable to obtain such goods, mimicry was the next best thing. One of the results of this was that embroidery was used to emulate the designs woven into silks, three good examples being the late 8th- to 9th-century Oseberg G, the late 9th- to 10th-century Llangorse textile and parts of the borders of the 11th-century Bayeux Tapestry.

---

22 A.E. Christensen and M. Nöcket, *Osebergfunnet: Bind IV Tekstilene* (Oslo: Museum of Cultural History, 2006), 265, 266, 390.

23 For discussions on this see: J. Blair, *The Church in Anglo-Saxon England* (Oxford: Oxford University Press, 2005), 8–51; N.J. Higham, 'From Tribal Chieftains to Christian Kings', in *The Anglo-Saxon World*, eds N.J. Higham and M.J. Ryan (London: Yale University Press), 126–178; B. Yorke, *The Conversion of Britain 600-800* (Harlow: Pearson Education, 2006), 122–133.

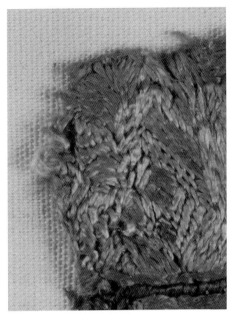

Figure 35. Detail of Oseberg G showing the stylised animal, © Museum of Cultural History, University of Oslo, Norway

Figure 36. Detail of tablet woven band 34D from Oseberg, showing a swastika cross, © Museum of Cultural History, University of Oslo, Norway

Oseberg G is embroidered with a cross-like motif. In between the arms of the cross are the partial remains of what Christensen and Nöckert consider to be stylised bird motifs, their heads turned backwards facing their bodies and one raised leg depicted (Fig. 35).[24] Christensen and Nöckert believe that the piece was intended to mimic a woven textile and point out that the style of the birds' heads is similar to those found on a tapestry-woven band from the same burial (24A). I have observed that the pattern is also in the same style as the swastika crosses on a tablet-woven band 34D (Fig. 36). Additionally, the birds' heads are located in the gap where the arms of the cross-like motif meet, forming a four-way repeating pattern, as seen on imported silks. This again indicates cross-fertilisation with woven textiles.

Turning to the Llangorse textile, Hero Granger-Taylor and Frances Pritchard have argued that the designer knew of the motifs woven into silk fabrics that were making their way to Anglo-Saxon England.[25] The incorporation of stylised motifs set out along both vertical and horizontal axes (and mirroring each other), points to designs from Byzantine and central Asian silks, and they highlight the similarity between the lions with their spots found on silks from Imperial Byzantine workshops and the lions with their three legs and spots on the Llangorse textile.[26] We can infer that the decoration of this piece was meant to imply the wearer's power was comparable to that of the emperor of Byzantium. So while the Welsh owner of this probable tunic was unable to gather enough silk together to make such

---

24 Christensen and Nöcket, *Osebergfunnet: Bind IV*, 400.
25 H. Granger-Taylor and F. Pritchard, 'A Fine Quality Insular Embroidery from Llangors Crannóg, near Brecon', in *Pattern and Purpose in Insular Art: proceedings of the fourth international conference on insular art held at the National Museum and Gallery, Cardiff 3–6 September 1998*, eds M. Redknap, N. Edwards, S. Youngs, A. Lane and J. Knight (Oxford: Oxbow Books, 2001), 91–99 (95–97).
26 Granger-Taylor and Pritchard, 'Fine Quality Insular Embroidery', 95–97.

a garment, he or she decided to mimic the status implied by the design through fine needlework, an available and presumably well-practised technique within the region.

Evidence demonstrates that the conversion of Anglo-Saxons to Christianity took place in the 7th century;[27] surviving embroideries associated with the new religion survive from a little later, the 7th to 8th centuries. These are Durham A and B, which are considered to be Christian because they were discovered with the stole, maniple and girdle/headdress in the tomb of St Cuthbert. Both Durham A and B are tablet-woven bands that possibly incorporate a form of embroidered wrapping stitch. Hero Granger-Taylor suggested that it represented an embroidered form of *latté* (this is where one or more weft *lats* – throws of the shuttle – change colour independently of the design) seen on silks from central Asia. This too, then, can be explained as embroiderers mimicking styles seen on other textiles: in this case rare ones (see. Pl. 13b, c). The embroidery's purpose is therefore to emulate not only the stitch and style but also the status and symbolism that such silks would hold within Anglo-Saxon society.

This is not the only case of mimicry in the Durham embroideries. I believe a second example can be seen on Durham D and E, the stole and maniple. Grace Crowfoot noted the similarity between the design on the vestments and the accompanying tablet-woven bands.[28] I believe this can be taken a step further. Nearly all surviving examples of couched gold-work in the cohort were worked in alternate bricking (a technique where the embroiderer offsets each row of couching stitches so the finished effect mimics a brick wall) including the backgrounds of the stole and maniple. However, the halos of the figures on both vestments were couched in more elaborate patterns which are similar to the brocading found on some of the tablet-woven bands associated with St Cuthbert, for example, the zig-zag effect on the halo of deacon Peter on the maniple, and the pattern on band 9, which is brocaded in silver-gilt thread (Pl. 18a, b). Although not exactly the same, the similarity between the brocading and the couched patterns can also be seen on the halo of Pope Gregory I on the maniple and the pattern on band 10 (Pl. 18c, d). This demonstrates the interchangeable nature of designs for textiles and the fluidity of use of artistic styles across different types of object. It is possible that the embroiderers who witnessed the patterns being created on objects such as the tablet-woven bands responded by choosing to incorporate similar versions into the embroideries.

Both these vestments are very much of the Christian world, not only because of their intended ecclesiastical function and later roles of gift to a saint and contact relics, but also in terms of design and use of materials. The motifs incorporated in the design were chosen with care and can be seen as part of a larger move within religious circles to use vestments to convey a many layered message. The first of these is political: the inscription stitched onto the reverse of the end tabs on both the

---

27 See Yorke, *Conversion of Britain*.
28 G.M. Crowfoot, 'The Braids', in *The Relics of St Cuthbert*, ed. C.F. Battiscombe (Oxford: Oxford University Press, 1956), 433–469, 433.

stole and maniple, tells us that Ælfflæd, queen to Edward the Elder, commissioned the vestments for Bishop Frithestan, bishop of Winchester from 909 to 931 (see Durham C–E) (Pl. 18e). As a political donation the vestments bestowed religious favour and influence on the commissioner. During this period women were not allowed to enter the sanctuary of a church, the place where the altar is situated and the mass said.[29] Nevertheless, by commissioning the work and having her name embroidered on the reverse of the stole and maniple, Ælfflæd was secretly placing herself at the altar during the most sacred rites of the Church, and arguably, closer to God. There is a second dimension to this gift. Vestments were seen not only as attire in which the priest enrobed himself for ritual, they also incorporated special meanings that helped the wearer live out his vocation more completely. With the development of more mystical ideas within the Church, vestments took on a level of symbolic or supernatural reverence;[30] for example, in c. 820 the writer Amalarius stated that the stole symbolised the 'light yoke of Christ' and the maniple, 'pure and pious thoughts wherewith we wipe away the disorders of the mind which arise from the infirmity of the body'.[31] Taking this a step further, I would argue that additionally the decoration incorporated onto the vestments took on linked meaning, and thus another layer of significance was incorporated into each vestment. On the Cuthbert vestments the use of prophets, popes and deacons in the designs helped to remind the priest of his role within the Church and how he should conduct himself.

At the same time women in elite circles were developing ways in which to exercise influence within the Church. They did so by patronising members of the clergy through gifts such as much needed vestments and other soft furnishings. Maureen Miller points out that these items were not easily accessible to the clergy because they did not live within secular household structures. As a result, clothing, vestments and soft furnishings that would usually have been produced or bought by women of the house were not, so the clergy were often reliant on gifts.[32] Elite women could therefore exert influence over members of the clergy, while they in turn could attempt to curry favour with the ruling elite. Thus, the Cuthbert embroideries were probably meant to place Ælfflæd closer to God but also intended to exert an influence at a more worldly level.

When the embroideries were taken to Durham by king Æthelstan in c. 934 their role changed. This event signals the beginning of the second life cycle of the embroideries. Instead of their continued prominent display in the sanctuary, they were repositioned hidden from view next to the body of Cuthbert, a powerful saint, whose relics lay at the centre of this important religious institution. At this time such establishments

---

29 M.C. Miller, *Clothing the Clergy: virtue and power in Medieval Europe c. 800–1200* (London: Cornell University Press, 2014), 145.
30 P. Johnstone, *High Fashion in the Church* (Leeds: Maney, 2002), 5.
31 C. Hohler, 'The Stole and Maniples (b) the iconography', in *Relics of St Cuthbert*, ed. Battiscombe, 396–408 (397).
32 For discussion about the political use and meaning of vestments see: Miller, *Clothing the Clergy*.

were entwined with the country's politics with many of those in holy orders being of high or royal lineage. The favour of giving such beautiful and expensive embroideries therefore carried both ways; the prayers of the monks and the power of the saint could, in the early medieval mind, influence the outcomes of Æthelstan's battle against the Picts. In turn political policies might be tweaked to favour the religious community, endowing it with greater power and influence. Thus, the Cuthbert embroideries today give scholars an insight into both the political and religious state of England during the early medieval period and the minds of those who inhabited it.

Phase 3

As the Anglo-Saxon kingdoms evolved, merged and cemented themselves into larger units, and eventually, a single kingdom, Anglo-Saxon identity evolved. Although outside influences continued to play a role, what we might think of as a specifically Anglo-Saxon artistic style emerged and this is reflected in some of the surviving embroideries. As we see with the Kempston fragment, particular artistic forms involving specific materials – metalwork, sculpture, manuscripts – may initially have been associated with particular territories, but as they became dispersed, they evolved to incorporate different motifs, materials and meanings, becoming an amalgamation of the original. For instance, the late 8th- to 9th-century Maaseik embroideries, show the influence of Byzantine textiles in their use of threads and pearls (Pl. 19a, b).[33]

In their analysis of the art historical aspects of the Maaseik embroideries, Mildred Budny and Dominic Tweddle argue that the motifs seen on the eight Maaseik embroideries are closely related to similar motifs found on southern English artwork of the late 8th to early 9th centuries.[34] They provide numerous examples of manuscripts, metalwork and stone carvings displaying similar decorative features, although they acknowledge that none contain all the motifs. Many of these instances, including manuscripts from Canterbury and the Priors Barton cross shaft from Winchester, are thought to have been produced in southern England, leading the authors to suggest the Maaseik embroideries were made there too. The evidence is not conclusive, however.

Leslie Webster has shown that both the arcade and roundel strips, with their interlace, and animal and bird motifs, are related to the art of 8th-century Mercia in central England, particularly that of Breedon. She goes on to argue that the most prominent motif of the style is the intertwined, biting animals based on designs found on silks and other goods brought to Anglo-Saxon England from the Byzantine Empire. She also observes that the entwining of the animals and their opposed stance harks back to earlier Anglo-Saxon artistic forms, creating a uniquely Mercian style.[35] It is

---

33 Webster, *Anglo-Saxon Art*, 114.
34 M. Budny and D. Tweddle, 'The Maaseik Embroideries', *Anglo-Saxon England*, 13 (1984), 65–96 (78–84).
35 Webster, *Anglo-Saxon Art*, 114, 138, 139.

therefore probable that the embroideries were created in a workshop either based in Mercia or influenced by this Mercian art form.

Elizabeth Coatsworth notes the striking similarities between the design of the Maaseik embroideries and that of Oseberg E (see Fig. 13), with its couched medallions depicting naturalistic animals.[36] The Maaseik roundels and creatures (Pl. 19c) are couched in gold while the possible lions/griffins on the Oseberg piece are worked in silk thread. However, the backgrounds of both embroideries are completely covered in silk thread worked in split stitch (see Glossary) while the areas between the roundels on both embroideries contain foliage motifs. The Maaseik piece uses geometric forms within the silk-work to create jewel-like patterns. Unfortunately, the Oseberg fragment is so faded and stained that the original colours are not clear, but it is reasonable to deduce it may have made use of similar techniques, and that both embroideries were influenced by the same artistic style.

Figure 37. Friskerton sword, Lincolnshire, Museums Sheffield

Oseberg E is not the only piece from this group that can be linked to Mercia. Oseberg H is also relevant. Christensen and Nöckert state that the geometric designs of the Oseberg silk embroideries have parallels in Anglo-Saxon art and manuscript illumination from the second half of the 8th century.[37] Geometric motifs are seen on manuscripts and metalwork dating to the 7th century and occasionally in the borders of carpet pages in 8th-century illuminated manuscripts attributed to the Canterbury Style, for example, the Vespasian Psalter. However, none show similarities to the design of Oseberg H.[38] Those examples that appear most reminiscent of the design on Oseberg H (see Fig. 16) can be seen on two 9th-century swords. The first was discovered in the River Witham at Fiskerton, Lincolnshire, and dates to the second half of the 9th century, while the second was found at Gilling Beck in North Yorkshire, and dates to the late 9th century. Both are decorated with a number of incised silver bands on the hilt grip (Figs 37 and 38). These designs are basic geometric forms inscribed within borders. The Fiskerton sword's design is most like that of Oseberg H. The silver band consists of four small diamonds, each containing a dot, within diamond shaped borders. Each row of diamonds lies point-to-point. In the space between border and outer edge of each band, additional, foliate or possibly geometric patterns are incised.

---

36 E. Coatsworth, 'Stitches in Time: establishing a history of Anglo-Saxon embroidery', in *Medieval Clothing and Textiles*, 1 (2005), 1–27 (9–11).
37 Christensen and Nöcket, *Osebergfunnet: Bind IV*, 401.
38 For examples see L. Webster and J. Backhouse, eds, *The Making of England: Anglo-Saxon art and culture AD 600-900* (London: British Museum Press, 1991), 27, 30, 50, 198, 199, 216.

Figure 38. Gilling Beck sword, North Yorkshire, image courtesy of York Museums Trust: http://yorkmuseumstrust.org.uk/: CC BY-SA 4.0

The design is not an exact reflection of that on the embroidery, but there are clear similarities: diamond shapes contained within a border.

The decoration on both swords is a late form of the Trewhiddle Style,[39] which was popular in Anglo-Saxon England between the late 8th and early 9th centuries. Webster argues that it was most commonplace in Mercian art and waned once the Mercian Kingdom lost its power in the early to mid-9th century, and Viking raids became more destabilising.[40] She suggests similar swords found in northern England and Norway is evidence of Norse activity in Anglo-Saxon England.[41] Movements of objects linked to Scandinavian raids could also account for the embroidered silk-work discovered in the Oseberg burial.

---

39 This artistic style is named after the find place of an Anglo-Saxon hoard discovered in Trewhiddle, Cornwall, in 1774.
40 Webster, *Anglo-Saxon Art*, 146.
41 L. Webster, 'Metalwork, wood and bone', in *Making of England*, eds Webster and Backhouse, 268–283 (277).

Figure 39. Fetter Lane sword hilt, London, © Trustees of the British Museum

Discussing Oseberg J (see Pl. 12), Christensen and Nöckert liken the embroidery's spiral design to that on the hilt of the Fetter Lane sword, discovered in London, which dates to the late 8th century.[42] It is silver inlaid with niello. On one side is a small spiral design encased in a stylised tendril motif surrounded by foliage patterns. On the other is a three-banded spiral which starts in the centre and swirls outwards (Fig. 39). Within each band are geometric and foliage motifs which are kept distinct from each other through the use of an outlining border. At the centre of the spiral is the head of a serpent. Webster points out that this design is typical of the Trewhiddle Style while Christensen and Nöckert connect the sword with the Mercian Kingdom through the serpent motif.[43] This is also true for the geometric shapes stitched on fragments Oseberg H and Oseberg I.

---

42 Christensen and Nöcket, *Osebergfunnet: Bind IV*, 332.
43 L. Webster, 'Metalwork', in *Making of England*, eds Webster and Backhouse, 220–239 (221–222); Christensen and Nöcket, *Osebergfunnet: Bind IV*, 332.

The link between the design of the Maaseik embroideries and the Oseberg fragments and the Trewhiddle Style suggests they may have been produced somewhere in the Mercian area and become dispersed from their origins over the course of their lives as they were gifted, appropriated or traded. As a result, this Anglo-Saxon art form spread to northern territories on the Continent, feeding the evolution of artistic styles.

With the decline of Mercian supremacy and the rise of the house of Wessex and the Danelaw in the 9th century, and the eventual coming together of the English kingdoms as one country, came the rise of new artistic styles. King Alfred (871–899) instigated a renaissance which influenced the refashioning of Anglo-Saxon identity in the face of the Viking onslaught, and new artistic styles evolved. Traits of these art forms can be used to help position and date the Milan fragment in its embroidery context.

Alberto de Capitani d'Arzago, the first to study the Milan fragment, dated it to the 11th century because the gold-work was found in association with dark blue Islamic silk 'S3', which is possible to date to that period (see catalogue entry).[44] However, as the catalogue entry shows, the combination of textiles that make up what is known as 'Dalmatic two' where the embroidery is found (a dalmatic being a wide-sleeved full-length loose vestment, see Glossary), date from the late Roman period through to the 19th century, so it is far from certain that the embroidery was created at the same time as the blue Islamic silk was woven.

In the 1950s Grace Crowfoot noted that the materials and technique were identical to Durham D and C although the design was larger and not so 'dainty' as those used on Durham C, leading her to suggest that the Milan embroidery was made in Anglo-Saxon England in the early 10th century. She went so far as to suggest that it may even have been made in the same workshop as the Durham embroideries.[45] But in 1984 Mildred Budny and Dominic Tweddle highlighted particular design elements that suggested to them an earlier, 9th-century, construction date. A scrolling stem with acanthus leaves, a bird viewed in profile, an insect, which they suggested was a bee, and a repeating stepped geometric pattern, indicated this earlier date. They pointed out that the upper guard of the 9th-century Abingdon Sword is decorated with a sub-rectangular frame containing a stylised insect seen from above (Fig. 40) and in the Book of Kells, the initial *chi* on 34r contains two moths seen from above and confronting each other (Pl. 20a). They also pointed out, however, that insects are rarely found on Anglo-Saxon or Hiberno-Saxon art and felt, therefore, that if this is indeed what it represents on the Milan fragment, it may indicate a significant dating feature.[46]

---

44 See G.M. Crowfoot, 'Note on a Fragment of Embroidery from the Basilica Ambrosiana in Milan', in *Relics of St Cuthbert*, ed. Battiscombe, 392–394 (392); H. Granger-Taylor, Unpublished extract. I am grateful to Frances Pritchard, formerly of The Whitworth for a copy of this work; A. de Capitani d'Arzago, *Antichi Tessuti della Basilica Ambrosiana*, new series 2 (Milan: Biblioteca de L'Arte, 1941), 41.
45 Crowfoot, 'Note on a Fragment from Basilica Ambrosiana', 393.
46 Budny and Tweddle, 'The Maaseik Embroideries', 86.

Figure 40. Detail of AN1890.14 the Abingdon Sword hilt, image © Ashmolean Museum, University of Oxford

In an unpublished discussion, Hero Granger-Taylor has argued that the poor condition of the Milan embroidery and the lack of comparative pieces means that its classification as English in origin should be treated with caution.[47] She points out that vine-scroll motifs were also popular in Italy, citing the borders of the second *Exultet* manuscript at Troia, Italy, and believes it is impossible to judge whether the piece was produced in the 10th or the first part of the 11th century. As Elizabeth Coatsworth has asked of this piece, and it is a question that can be asked more generally, should technique or design elements be given more weight when dating embroidery?[48]

My own analysis has highlighted a number of issues with regard to dating the Milan embroidery. As documentary sources and the other surviving examples in this catalogue show (Maaseik A, B and C), the technique of couched gold-work was being expertly deployed in Anglo-Saxon England from at least the 8th century, pointing towards its development as a technique considerably earlier. Extant archaeological examples demonstrate that gold-work was being produced in mainland Europe, particularly in the Frankish Empire, from at least the 7th century.[49]

Since it has not been possible to view the Durham embroideries due to their ongoing inaccessibility to researchers, it is difficult to compare and comment on similarities between Durham C and the Milan piece. It is possible, however, to compare a microscopic image of the Ushaw fragment, which was originally part of Durham D (Pl. 20b, c). When seen side-by-side it is striking how similar in style the working methods used for the two pieces are. This is in contrast to the workmanship of a comparative section of the earlier Maaseik embroideries (Pl. 20d). The technical handling of the Milan gold and stitch work is much superior to the surviving section

---

47 Granger-Taylor, 'Unpublished Report'.
48 Coatsworth, 'Stitches in Time', 19.
49 Coatsworth, 'Stitches in Time', 4.

Figure 41. Colyton cross shaft, Devon, © Corpus of Anglo-Saxon Stone Sculpture, photographers K. Jukes and D. Craig

of the roundel from Maaseik, and is comparable to that of the Ushaw fragment. The evidence is strengthened here by the similarity of the materials. The gold threads used to create both the Durham embroideries C, D and E and Milan have a red core,[50] and seem to have been wound by individuals with similar levels of dexterity. The reverse of Durham C shows the vine scroll motif has been worked in gold thread couched with a red silk thread and outlined in stem stitch with a thicker red silk thread in a similar manner to the Milan motifs (see Pl. 4b). The couching patterns, including the veins of the leaves in a darker thread, are also visible on Durham D and E. Technically, then, the evidence points to the Milan fragment having been produced during the 10th century and in a professional workshop that used strikingly similar standards and methods of construction, and possibly the same design exemplars.

The design is more difficult to date, however. The foliage has some similar elements to the leaves embroidered on the reverse of Durham C, which is characteristic of the Winchester Style (see Pl. 4a, b). However, the Durham embroidery is much more rigid in its design and layout than the Milan piece (Pl. 21a, b), even when the 1940 repositioning and remounting of the Milan pieces is taken into consideration. Although Granger-Taylor demonstrated that there are southern Italian manuscripts that contain similar border patterns to the Milan embroidery, such scrolling foliage can also be seen on the dedication page of Bede's *Life of St Cuthbert* (Pl. 22), which was made in Winchester, and is an example of early period Winchester Style dating to before 937. This example shows the foliage contained within a border of gold similar to the Milan piece. Within the border, gold vines scroll around each other and end in leaves, some of which sprout flowers, more leaves, trefoils and berries. Although not exactly corresponding to the scrolling vegetation seen on the Milan embroidery, it is very similar.

An example of early 10th-century sculpture, again made in the Winchester Style, also exhibits a similar pattern. The cross shaft now housed in St Andrew's church in Colyton, Devon (Fig. 41),[51] shows a vine scroll, albeit inhabited with animals, that

---

50 E. Plenderlieth, 'The Stole and Maniples (a) the technique', in *Relics of St Cuthbert*, ed. Battiscombe, 375–396 (377). Crowfoot, 'Note on a Fragment from Basilica Ambrosiana', 393.

51 Rosemary Cramp, *Corpus of Anglo-Saxon Stone Sculpture volume 7: South-West England* (Oxford: Oxford University Press, 2006), pp. 80–82, pl. 3.

Figure 42. Detail of the central motif showing a bird inhabiting a vine scroll: Colyton cross shaft, Devon, © Corpus of Anglo-Saxon Stone Sculpture, photographers K. Jukes and D. Craig

Figure 43. The 7th-century gold and garnet cicade from Horncastle, Lincolnshire, © Lincolnshire Archives

contains leaves and buds that are very like those embroidered on the Milan piece. It, too, is contained within a border made of straight bands. In the inside bottom corner of the border is a leaf bent at right angles, which is repeated in mirror form at the top of the plants carved into the base. This is reminiscent of the stepped geometric or foliate shapes we see in the corners of the squares on the Milan fragment.

The bird viewed in profile has almost an exact parallel on the Colyton cross shaft (Fig. 42): the neck of the Milan bird is not as elongated, and the body is slightly dumpy when compared to the sculpted version (see Pl. 21a), but the similarities are unmistakable. Where the features of the sculpted bird have been picked out with indented lines, on the embroidery they have been accented with embroidered lines.

The possible insect is more difficult to relate to Anglo-Saxon art. The two examples given by Budny and Tweddle show little similarity in terms of form or decoration. Examples of bees viewed from above can be found on metalwork originating across Europe and throughout the early medieval period, but none look the same as the Milan exemplar (see for example Fig. 43). The motif may feasibly be interpreted instead as foliage, and a sculpted comparison can be seen at the bottom of the narrow side on the upper shaft of a cross at Todber, Dorset (Fig. 44). Inside the carved frame is an acanthus scroll stem and leaves that terminate at the top in a bunch of grapes. The stem rises from each side of the frame and it is this section that looks like an inverted and split form of the Milan motif. The carving also sits within a frame. The cross shaft is sculpted in the Winchester Style and dates to between the late 10th and early 11th centuries.[52]

As so little of the original embroidery and this motif survive, its actual design is still somewhat of a mystery. From the evidence available, it can be argued that the embroidered design was

---

52 Ibid., pp. 114–116, pl. 109.

Figure 44. The upper cross haft from Todber, Devon, © Corpus of Anglo-Saxon Stone Sculpture, photographers K. Jukes and D. Craig

probably a framed vine scroll inhabited with birds, and possibly other creatures. It may have been a form of the Tree of Life, as seen on the Colyton cross shaft. The foliage design, the bird, and to some extent the unknown motif and embroidery technique of the Milan fragment, point to a 10th-century creation date in a workshop that specialised in the Winchester Style. The surviving embroidery was most likely part of a larger piece that was cut down and recycled when it was attached to the dalmatic in the 11th century, giving it a new incarnation, or it could have been part of an embroidered band that was originally applied to another garment and then recycled when it was attached to the dalmatic – in a similar manner to the Maaseik embroideries, which were later applied to the composite piece now called the *casula* of Sts Harlindis and Relindis.

During times of peace, prosperity and/or renewal there can be an expansion in creative arts. Martin Ryan has pointed out that despite the political problems encountered during the reigns of Æthelred (d. 1016) and Cnut (d. 1035) this was a period of intellectual renewal and cultural accomplishments within creative circles.[53] Dodwell has shown that during the later part of the Anglo-Saxon era, wealth and rank were demonstrated through the quality of garments people were wearing and the accessories they adorned themselves with.[54] It may be that these expressions of wealth, status and renewal can be seen in the survival of a new type of embroidery from this later period – decorated footwear. All the extant examples of embroidered vamps and top bands (see Glossary) from this timeframe may be due to the type of conditions needed in order for leather and wool to survive but as these conditions do exist elsewhere in the country it is interesting that this form of embroidery has not appeared on earlier pieces. There are two 7th-century examples of shoes/slippers that were probably decorated with embroidery from the Ship Burial in Mound 1 at Sutton Hoo (see Table 3), but these would have been made for royalty and it is safe to assume that, at this date, such decoration was not common

---

53 Martin J. Ryan, 'The Age of Æthelred', in *Anglo-Saxon World*, ed. Higham and Ryan, pp. 335–373 (p. 352).
54 C.R. Dodwell, *Anglo-Saxon Art: a new perspective* (Manchester: Manchester University Press, 1982), 174.

for the majority of the population. Indeed, Gale Owen-Crocker has pointed out that going barefoot may have been more common than is supposed.[55]

The earliest example of embroidered footwear comes from Winchester and dates to the early 10th century. Winchester was at this time a centre of ecclesiastical learning and a seat of power, so it is not unlikely that many of those within the town's elite circles wore decorated garments. The surviving embroidery consists of rows of stitching which form a vamp stripe that runs down the centre front of the shoe or boot. As with all the examples cited here, the stitch holes only pierce the grain side of the leather, thus creating a tunnel stitch hole that did not break the leather's surface on the reverse (see Glossary). The rows lie 3.5 mm apart and created a stitch length of 2 mm. This stitching appears to be slightly larger than the other extant examples. Although there is embroidery thread in the holes, no complete stitches have survived. It is possible that they formed a raised plait stitch similar to that of the shoe at Coventry (see Pl. 7a) which dates to between the late 11th and early 12th centuries (see above). Equally, the embroidery could have consisted of simple lines of back stitch or running stitch (see Glossary).[56] There are two other decorated vamps, both from London (B and C) and both date to between the 1070s and 1080s. Both examples survive as fragmented pieces of leather that appear to have been torn and cut ready for re-use. Both are also decorated with four rows of fine stitch holes that create one vamp stripe. London B has silk thread extant in some of the holes while the thread type of London C has not been discussed in the published volumes. Unfortunately, there are no images of these two shoes in the published work and I was unable to view them in person. As a result, I have not been able to determine the type of stitch that was used to decorate either vamp.

The final embroidered piece is partially preserved top band that was also found in London (A). This object was decorated with wool threads threaded through two parallel rows of holes, measuring 1 mm long by 1 to 2 mm apart (see catalogue). This seems to be a standard size and distance for such holes.[57] The surviving evidence suggests that the two rows of holes were decorated with two parallel rows of running stitch (see catalogue entry).

It is difficult to draw any overarching conclusions from the surviving footwear evidence. London B and C were both found in an open area within the Guildhall site. Excavations of the area brought to light evidence that it had, during this period, been a place occupied by craftsmen working predominantly as shoemakers and cobblers,[58]

---

55 G.R. Owen-Crocker, *Dress in Anglo-Saxon England, revised and enlarged* (Woodbridge: Boydell Press), 123.
56 See M. Volken, *Archaeological Footwear: development of shoe patterns and styles from Prehistory till the 1600's* (Zwolle: Spa-Uitgevers, 2014), 198.
57 These measurements were taken during a research trip to the Museum of London Archive in 2015. During the same trip I measured the stitch holes of a number of examples of footwear which gave similar sizes.

and it is probable that the cut down vamps of London B and C were part of this process. London A came from Milk Street which Alan Vince has shown is not as easy to define as the Guildhall. Vince has suggested that finds from Milk Street and the vicinity indicate occupation sites as he points out that there is not a sufficient amount of evidence to suggest any form of industry. Examples of silk have also been found at the site,[59] which suggests that it may have been a place of occupation for people who liked to express themselves through their clothing, as Dodwell highlighted.

Conclusion

Embroidery can be used in conjunction with other archaeological evidence to track and interrogate the changing terrain of Anglo-Saxon England. A combination of external and internal influences helped shape the early medieval period. Interchange with those outside the Anglo-Saxon world occurred through trading networks, gift exchange, raids and religious conversion. Other forces for change came from within, in particular, the renaissance brought about by Alfred in the process of reshaping a ravaged set of kingdoms into a renewed, unified country. Towards the end of the period those of elite groups began to express their power, wealth and status through the wearing of fine textiles and adornments. Interrogating the surviving fragments of early medieval embroidery allows us to envisage the ways in which material culture played a part in these processes.

The embroideries I have categorised as Phase 1, those influenced by the religion, politics and society of the northern zone, are many-layered, rather than simply decorative. Design may be multi-faceted, representing both ancestral and religious affiliation, and/or protective and functional style. Image and stitch were used not only as decoration, but as the means by which a form of symbolic protection was provided at seams and the edges of garments, or, with regard to the entwined beast/knot motif, around the whole object. The pagan peoples who stitched such embroideries thought of them as a way to ward off evil and provide protection against unknown threats in the wider world as embodied in their belief systems.

By the 10th century, belief in the protection of Germanic mythological creatures had been replaced by belief in a Christian God and veneration of prophets, popes, deacons and saints, and emphasis on the prayers of the Christian faithful. One set of theological beliefs had morphed into another, but use of beautifully worked

---

58 D. Bowsher, T. Dyson, N. Holder and I. Howell, eds, *The London Guildhall: an archaeological history of a neighbourhood from early medieval to modern times part 1* (London: Museum of London Archaeology, 2007), 19.
59 A. Vince, 'The Development of Saxon London', in *Aspects of Saxon-Norman London: II Finds and Environmental Evidence*, ed. A. Vince (London: London and Sussex Archaeological Society, 1991), 409–435 (430).

embroidery to express such ideas remained the same. To keep it safe and near to one's person was a means of ensuring its influence.

Towards the end of the timeframe embroideries took on new traits as a more complex Anglo-Saxon identity developed. Society had evolved from a group of tribal kingdoms with pagan beliefs to a nation with its place in the Christian world. Embroidery had developed from entwined mythical beasts protecting and reinforcing seams to elaborate pictorial images involving metal thread and silk-work. Such works were used to provide honour to saints and/or give favour and protection to those who could afford it. In elite circles – both ecclesiastical and secular – wealth, status and power were expressed by the wearing of good quality textiles and adornments, and these included items decorated with embroidery.

Chapter 5

Early medieval embroidery production in the British Isles

Introduction

The production of embroidery in the British Isles in the early medieval period, together with the lives, training and working conditions of those who produced it, has hitherto proved inaccessible. There are three main reasons: the paucity of documentary sources discussing its production (particularly beyond elite circles); the lack of archaeological evidence, and, perhaps following on from this, little research effort dedicated to the project of piecing together potential sources of information.

Nevertheless, in the introduction to *English Medieval Embroidery*, which was, at the time, a defining publication on embroidery, Grace Christie demonstrated the way in which painstaking analysis of embroidery materials and techniques might also be deployed to provide useful evidence concerning production.[1] This involved examining individual extant pieces in order to identify the ground fabric (see Glossary) on which the stitching was worked, and the threads and other materials used to create embroidery stitches and applied components (such as tablet-woven bands and fringes). However, of the two sets of major early medieval embroideries that she believed to have been produced in the British Isles (the Cuthbert embroideries and the Bayeux Tapestry), Christie only studied the Cuthbert examples (Durham C, D and E) in detail, and then only by eye. However, she knew of early scientific analysis undertaken by Professor Littlejohn of Edinburgh University using comparative microscopical, micrometrical and chemical testing, and an analytical examination by Professors Laurie and Gibson at the Heriot-Watt Technical College, Edinburgh, resulting in the identification of the embroidery threads and the core of the gold thread as

---

1 [Mrs] A.G.I. Christie, *English Medieval Embroidery: a brief survey of English embroidery dating from the beginning of the tenth century until the end of the fourteenth* (Oxford: Clarendon Press, 1938), 1–44.

silk.[2] Other embroideries have come to light since Christie's period and their more recent discovery has provided impetus for more advanced scientific study, such as microscopic analysis of the fibres and other materials used in their construction. The embroidered sleeve wrist-edging from York (York B) discovered during the excavation of 28–29 High Ousegate, York, is a good example of a recent find that has benefited our understanding by the application of such techniques.[3]

Art historical perspectives have also been brought to bear on embroideries involving iconographic imagery, and sometimes allow us to link an embroidery to stylistic forms found elsewhere: in manuscript art, for instance. The much studied stole and maniple from the tomb of St Cuthbert in Durham (Durham D and E) is just such an example. Stylistic elements from both embroideries match manuscripts in the Winchester Style, which originated in Wessex. Additionally, in the case of the Cuthbert stole and maniple a timeframe for production can be derived from the embroidered donor-receiver inscriptions, where both donor and recipient are identifiable people: a historical queen, Ælfflæd (d. before 916), and a bishop, Fristhestan (909–931). This provides us with approximate dates for production, AD 909–916. Here, iconography and inscriptions provide mutually reinforcing evidence.[4] Not all embroideries yield such rich detail – indeed the incidence of donor information and iconographical evidence here is something of an exception. Nevertheless Christie's work, York B and the art-historical work on Durham D and E position individual pieces as valuable sources of information, and the second half of this chapter seeks to use the cohort of early medieval embroideries to build on this work. However, the chapter begins by reassessing published historical and archaeological material through the lens of the embroidery-practitioner. While there is little in the way of clear archaeological evidence that points specifically to either embroidery production or embroiderers' working conditions, and documentary evidence is also slight, there is potential for fresh interpretation. When existing historical sources are used in conjunction with archaeological evidence associated with textile production in general (as opposed to embroidery in particular), they provide valuable context for the workers and their lives. These two largely untapped resources can be used to generate a framework for production.

---

2 G. Baldwin Brown and [Mrs] A. Christie, 'S. Cuthbert's Stole and Maniple at Durham', *The Burlington Magazine*, 23 (April 1913), 2–7, 9–11, 17 (3, 5); E. Coatsworth, '"A formidable undertaking": Mrs A.G.I. Christie and English Medieval Embroidery', *Medieval Clothing and Textiles*, 10 (2014), 165–193 (186).

3 P. Walton Rogers, 'Textiles, Cords, Animal Fibres and Human Hair', in *28-9 High Ousegate, York, UK*, eds N. Macnab and J. McComish (York: York Archaeological Trust, 2018), 14–41.

4 C. Hohler, 'The Stole and Maniples (b) the iconography', 394–409; E. Plenderlieth, 'The Stole and Maniples (a) the technique', 375–396; R. Freyhan, 'The Stole and Maniples (c) the place of the stole and maniple in Anglo-Saxon art of the tenth century', in *The Relics of St Cuthbert*, ed. C.F. Battiscombe (Oxford: Oxford University Press, 1956), 375–396 (409–432).

Reviewing the historical and archaeological evidence

Two now famous examples of professional embroiderers from Domesday Book, Aelfgyth and Leofgyth (also known as Leofgeat and Leviet), are regularly named by scholars, but few others are cited.[5] Ælfgyth held two hides of land in her own right in the pre-conquest period and half a hide from King Edward's household revenue, which was given to her by Godric, the Sheriff of Buckinghamshire. He stipulated that Ælfgyth could hold this land for as long as he remained sheriff, and she taught his daughter to embroider with gold thread, 'Aluuid puella. II. hid. quas potuit dare uendede⁷ cui uoluit. de dñica firma regis. E. habuit ipfa dim⁷hidá. quá Godric⁹ uicecom⁷ ei ćceffit quádiu uicecom⁷. e⁷e⁷t.ut illa doceret filiá ej⁹ Aurifrifiū opari'.[6] We do not know whether Ælfgyth ran a school or workshop, or worked alone and had a private arrangement with the sheriff. It would seem likely that she was granted this land as *lænland*, a form of loan or lease that lasted as long as Godric held office, after which it would revert back to the king.[7] The second woman, Leofgyth, held Knook in Wiltshire. Before the Conquest this land was held by Leofgyth's husband. This may be another example of *lænland*, which was leased for a set period, often for 'three lives', meaning that it could pass from husband to widow to child.[8]

Sources that contain recurring mentions of embroidery-workers from elite and royal circles focus on how worthy, religious and accomplished these women were but rarely discuss the more mundane topic of the production methods employed in their work. It is useful, therefore, to review the type of equipment and working conditions the workers would have needed to produce both domestic and high-status pieces from another angle. Drawing on experience – training and working at the Royal School of Needlework, it is possible to assemble a working profile of the environment in which early medieval embroiderers might have worked.

Equipment has not changed radically since at least the 15th century, from which period surviving documentary and illustrative sources show workers seated at wooden slate frames that hold taut the embroideries on which they are working.[9] The frames are set horizontally or at a slight angle on wooden trestles in order to allow the worker access to different parts of the needlework from both the front and the reverse. As Kay Staniland has pointed out, it is probable that the use of frames went back a number of centuries,[10] but because both the frames and the trestles were

---

5 E. Teague and V. Sankaran, trans., '13: Buckinghamshire', in *Domesday Book*, ed. J. Morris (Chichester: Phillimore, 1978), 149b, c; C. Thorn and F. Thorn, trans., '6: Wiltshire', in *Domesday Book*, ed. J. Morris (Chichester: Phillimore, 1979), 74b.
6 Teague and Sankaran, '13: Buckinghamshire', 149b, c.
7 For discussion of *lænland* see S. Baxter and J. Blair, 'Land Tenure and Royal Patronage in the Early English Kingdom: a model and a case study', *Anglo-Norman Studies*, 28 (2006), 19–46 (20).
8 Baxter and Blair, 'Land Tenure', 23.
9 See for example, a detail from Francesco del Cossa's fresco 'Allergory of March: Triumph of Minerva' in K. Staniland, *Medieval Craftsmen: Embroiderers*, 7th edn (London: British Museum Press, 2007), 49.
10 Staniland, *Medieval Craftsmen*, 27, 28, 32.

Figure 45. Line drawing of an embroiderer from Exodus XXXVI, 8-19, The Byzantine Octateuch, © Alexandra Lester-Makin

most likely made of wood (as they still are today), they have not survived in the archaeological record. It is therefore likely that similar frames and trestles were being used during the early medieval period, particularly during the later part when the production of silk-work and other forms of high-status and ecclesiastical embroidery was at its height. In order for these pieces to be produced successfully, they would need to be kept taut so that they did not warp during the production process. The earliest evidence for an embroidery frame is an illustration in the *Byzantine Octateuch*, which dates to the 11th century (Fig. 45). The image is associated with the Book of Exodus and the creation of the tabernacle. It shows a man leaning, somewhat uncomfortably, over a frame that has some sort of fabric attached to it. The man is seated on a stool, while the frame leans against a piece of furniture, possibly a bureau of some kind. Below him are examples of decorated cloth from the tabernacle and what must be the robes to be worn by the high priest or priests. Researchers have always assumed that this man is weaving because of the picture's association with the construction of the tabernacle, but the information given in Exodus about the creation of the soft furnishings – for both the tabernacle and the priestly robes – includes reference to them being decorated and embroidered.[11]

The way the fabric in the image is attached to the frame, and the manner in which the man is leaning over it with both hands on top of the work, with his right hand positioned as if it is holding something that is piercing the fabric, leads one to suspect that the image is actually depicting an embroiderer at work, or that the artist used a memory of an embroiderer as his inspiration. Indeed, if this image is compared to Fig. 46, which shows the author working at her frame, except for the placement of the hands and the fact that the author's frame is on trestles, the position and way the fabric is framed-up (attached to the frame), are almost identical.[12] The likeness between the two examples is even more obvious in a second photograph (Fig. 47). In this instance, both the author's hands are on top of the ground fabric re-threading a needle in a similar manner to that of the man in the biblical image. If the image from the *Octateuch* is a man embroidering, it is the earliest picture of an embroiderer known to the author. That it dates to the 11th century shows that frames were known

---

11 DRBO. ORG, 'Book of Exodus', in *Douay-Rheims Bible + Challoner Notes* (2001–2016) http://www.drbo.org/chapter/02001.htm (Accessed: 3 June 2016). See chapter 26, vs. 1–14, 31–37; chapter 28, vs. 31–35, 40–43; chapter 36, vs. 8–19, 35–38; chapter 39, vs. 1–32.
12 Fabric at the top and bottom of the frame is protected with tissue paper.

Figure 46. The author at her embroidery frame, using a slate frame resting on trestles, © Alexandra Lester-Makin

Figure 47. The author at her embroidery frame, both hands are positioned above the frame as she threads a needle, © Alexandra Lester-Makin

and used long before the 15th century. It is not unlikely that such frames were also being utilised in early medieval England.

Other basic equipment required by an embroidery-worker are needles and scissors. Both are found in the archaeological record but none, as yet, have been classified as embroidery tools. Needles have been discovered in burials and urban contexts and have always been assumed to be associated with functional sewing and leather-work,[13] although it is possible that they were used for embroidery. Needles and needle-cases have been found at sites as far apart as the cemetery at Kingston Down, in south-west Kent, which dates to the late 7th century and the Viking ship burial at Scar on the Isles of Orkney, which dates to between AD 875 and 950. Some needle-cases were found containing needles: for example, the tubular copper-alloy needle-case from Kingston Down contained two very fine needles of gilt copper-alloy, which could have been used for fine needlework such as silk embroidery. Other, open-ended needle-cases, such as those made of bone found in Viking Scotland, have needles that were inserted into a small piece of cloth which was then placed in the case, to keep them secure.[14] A fine, curved tool, possibly a needle, was also found at the Six Dials site in

---

13 P. Walton Rogers, *Textile Production at 16-22 Coppergate*, The Archaeology of York 17/11 (York: Council for British Archaeology, 1997), 1785; G.R. Owen-Crocker, *Dress in Anglo-Saxon England, revised and enlarged* (Woodbridge: Boydell Press, 2004), 277. For needles and shears see D.A. Hinton, *Southampton Finds volume 2: the gold, silver and other non-ferrous alloy objects from Hamwic* (Stroud: Alan Sutton Publishing, 1996), 48–50; P. Walton Rogers, *Cloth and Clothing in Early Anglo-Saxon England: AD 450–700* (York: Council for British Archaeology, 2007), 40–45. For shears also see P. Ottaway and C.A. Morris, 'The Leatherworking Tools Recovered', in *Leather and Leatherworking in Anglo-Scandinavian and Medieval York*, eds Q. Mould, I. Carlisle and E. Cameron, 3235–3240.
14 A. Smith, 'The Needle Tidy', in *Scar: A Viking Boat Burial on Sanday, Orkney*, eds O. Owen and M. Dalland (East Linton: Tuckwell Press in association with Historic Scotland, 1999), 95–6; Walton Rogers, *Cloth and Clothing*, 40–41.

Southampton. It was made of copper-alloy with two holes pierced through it at one end and a sharp point at the other, and may have had a similar use to the copper-alloy needles found at Kingston Down.[15] Today professional embroiderers use small curved needles in gold-work in order to sew the ends of gold thread to the reverse side of embroidery.

Many of the graves that have produced needle-cases and needles also contain shears of varying sizes, some of which are very small. A symbolic function has been suggested for these, because they have been found with other possible symbolic artefacts.[16] However, some of the smaller shears may have been functional, including a pair found at the cemetery at Mill Hill in Kent. These measure 90 mm with blades of 38 mm and are exactly the same size as the author's embroidery scissors: we might thus hypothesise that some of the smaller shears were functional and were used for embroidery.

As well as requiring specific equipment, embroiderers need particular conditions in which to work. It is likely that the type of working environment available to embroiderers would have changed over the six centuries that encompass the early medieval period. This era saw a highly dynamic society evolve, from a pattern of small settlements and farming communities into complex, regulated communities with urban centres and a more sophisticated agricultural system. Although there is no archaeological or documentary evidence that specifically points to embroidery workrooms or workshops, it is reasonable to hypothesise that they developed over time, and we can explore existing archaeological and documentary sources with this in mind in order to find reference to work spaces that would have been suitable for embroidery production. Ideally, embroiderers need access to good clear light, which many buildings during the early medieval period did not provide, and to clean air and surroundings to avoid dirt getting caught in the embroidery. This second requirement is particularly pertinent during the later part of the period when silk- and gold-work were being produced for elite circles and the Church. These types of embroidery suffer to a greater extent if contaminated by dirt and sweat.

It is useful to look at structures we know to have been utilised for textile production. Since embroidery is closely linked to the wider textile manufacturing processes, it is probable that during the early part of the period at least, women embroidered within the same working environment. In her 2007 publication, Penelope Walton Rogers has usefully demonstrated that during the 5th to 8th centuries in communities such as West Stow in west Suffolk and Mucking in south-east Essex, particular stages of textile production were done at set times of the year, making use of differing parts of the land and settlement. So while spinning and weaving may have taken place in

---

15 Hinton, *Southampton Finds volume 2*, 48–50.
16 Walton Rogers, *Cloth and Clothing*, 40. For a discussion on the possible symbolic function of chatelaines and objects attached to them see H. Geake, *The Use of Grave Goods in Conversion-Period England, c. 600–c.850*, British Archaeological Reports, British Series 261 (Oxford: Archaeopress, 1997), 57–58.

Grubenhäuser (sunken featured buildings), other tasks, such as wool combing, would have been carried out outside after the sheep had been sheared in early summer and the fleece had been sorted and washed; and others still may have been integrated (both spatially and in terms of time spent) alongside other chores.[17] Applying these perspectives, we may reason that the embroidery of clothes and other portable items could take place anywhere within the settlement, if sunny outside or, if the weather was inclement, within living or weaving quarters, and at any time of year, fitting around other more seasonal tasks. However, large embroideries would have to be attached to frames which would have been cumbersome to move, and more limiting.

The function of the *Grubenhäuser* itself has come under further scrutiny since the publication of Walton Roger's book. Some have argued they were actually storage buildings, as some of those discovered appeared to function as a cache for loom weights; others suggest the variety of material dug up is simply backfill dumped there once the building had fallen out of functional use.[18] However, it has also been suggested that larger *Grubenhäuser* were the centres of organised textile workshops and that the large number of loom weights found on the sunken floor (greater in number than would normally be found attached to a loom), can be attributed to the fact that wider looms may have been utilised to make larger textiles.[19] Nevertheless, it is unlikely that these buildings were associated with embroidery production because a lack of large windows would have meant they would have been too dark. This is also true of other types of building found on these early sites. A more promising alternative is offered by an early 8th-century example of metalworkers using dedicated outdoor work spaces at Moynag Lough in County Meath, Ireland. There are post-holes that probably held a wind-break or shelter associated with an outdoor heating area.[20] It is feasible that embroidery-workers had similar, possibly portable, outdoor structures, which gave shelter without blocking access to light. One possible example of such a structure can be seen in the 12th-century *Utrecht Psalter*. Here women are spinning and weaving in the fields under a canopy (Fig. 48). The artist may have seen similar scenes

---

17 Walton Rogers, *Cloth and Clothing*, 9–10, 41, 44; P. Walton Rogers, 'Cloth, Clothing and Anglo-Saxon Women', in A *Stitch in Time: Essays in Honour of Lise Bender Jørgensen*, eds S. Bergerbrant and S.H. Fossøy (Gothenburg: Gothenburg University, Department of Historical Studies, 2014), 253–280 (256).
18 H. Hamerow, *Rural Settlements and Society in Anglo-Saxon England* (Oxford: Oxford University Press, 2012), 62–64; G. Speed, *Towns in the Dark? Urban Transformations from Late Roman Britain to Anglo-Saxon England* (Oxford: Archaeopress, 2014), 131–132. Also see J. Tipper, *The Grubenhaus in Anglo-Saxon England: An analysis and interpretation of the evidence from a most distinctive building type* (Yedingham: Landscape Research Centre, 2004).
19 S.J. Plunkett, 'The Anglo-Saxon Loom from Pakenham, Suffolk', *Proceedings of the Suffolk Institute of Archaeology and History*, 34 (1999), 277–298 (295).
20 J. Bradley, 'Moynagh Lough: an Insular Workshop of the Second or Third Quarter of the Eighth Century', in *The Age of Migrating Ideas*, eds R.M. Spearman and J. Higgitt (Edinburgh: National Museums of Scotland, 1993), 74–81 (79).

Figure 48. Detail of women spinning and weaving, Utrecht Psalter. Utrecht, University Library, MS 32 folio 84r

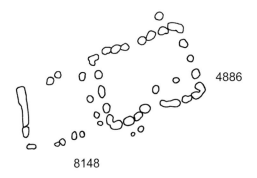

Figure 49. Line drawing of structures 4886 and 8148, Brandon, Suffolk, © Alexandra Lester-Makin, after Tester et al. (2014)

in the real world and used them as a basis for this illustration. It is certainly conceivable that such structures were used by embroiderers as well as those undertaking other textile work.

Another edifice with potential as a site of embroidery production, dating to between the mid-8th and 9th centuries, was excavated at Brandon in Suffolk. Located in an area of buildings associated with textile production, and overlooking the bleaching and dyeing section was structure 4886 (Fig. 49). It has been defined as possibly a craft workshop or, owing to the quality of its construction, which consists of regularly placed timbers and a clay floor, perhaps a dwelling used by a high-status person. The structure measured 20 m^2 and had inset entranceways. Adjoining it was structure 8148, which has been described as possibly a shed.[21] From the associated diagram, structure 8148 appears to have been approximately 15 m^2 (see again Fig. 49). This would make it a large shed. Its timber posts are not as regularly laid out as those of 4886, with space for what appears to be four entrances, one in each corner. It is possible that 8148 was a space that allowed in light whilst also providing shelter. Tantalisingly, then, this is the sort of structure that would be suitable for craftwork such as embroidery. As the following section explores in greater detail, embroiderers were attributed high social status and so it is conceivable that this structure would have been sited next to the living quarters of a person of status (structure 4886).

From the 7th century the development of larger sites or estates, known by archaeologists as 'central locations' (or central places),[22] established at a time

---

21 A. Tester, S. Anderson, I. Riddler and R. Carr, *Staunch Meadow, Brandon, Suffolk: a high status Middle Saxon settlement on the fen edge*, East Anglian Archaeology 151, ed. K. Wade (Bury St Edmunds: Suffolk County Council Archaeology Service, 2014), 14, 23, 361.

22 K. Ulmschneider, 'Settlement Hierarchy', in *The Oxford Handbook of Anglo-Saxon Archaeology*, eds H. Hamerow, D.A. Hinton and S. Crawford, 156–171 (161).

when social elites and the Church began to control production processes and the organisation of the land, seems to have partially centralised a number of activities including agriculture and craftwork. This led in turn to a level of specialisation in the production of goods, including textiles, and possibly, therefore, embroidery. Such development was partially linked to the evolution of overseas trade networks through which embroidered goods would surely have been exchanged.[23] The multi-functional purpose of these estates, producing a surplus for trade,[24] makes the presence of some form of embroidery workshop producing decorated items a likely proposition.

As Maren Clegg Hyer and Gale Owen-Crocker have suggested, an estate or workshop would also have greater access to raw materials because it would have the space and funds to enable it to produce, collect and store more than those producers of limited means.[25] So while small-scale domestic producers continued to work as they had always done, producing textiles for themselves and the extended family, a more organised textile production system may have been developing within larger settlements and estates, as the excavated examples of Flixborough in Lincolnshire and Brandon in Suffolk demonstrate.[26]

This aspect of estate development also provides a favourable context for the development of more sophisticated textile decoration, and an associated evolution of embroidery workshops. Indeed, there is a hint of this in what is possibly the earliest recorded evidence for a named embroiderer. The earliest documented example of the sort of estate we are discussing is in a charter dating to the early 9th century. In it is recorded that Eanswitha was granted a life-long lease for a two-hundred-acre farm at Hereford by Denewulf, Bishop of Worcester. This was in payment 'that she [Eanswitha] should continue to renovate, clean and add to

---

[23] For discussions on the re-organisation of production and development of trade networks see: J. Naylor, *An Archaeology of Trade in Middle Saxon England*, British Archaeological Reports, British Series 376 (Oxford: Archaeopress, 2004).

[24] Walton Rogers, *Cloth and Clothing*, 44, 47; D. Powlesland, 'Early Anglo-Saxon Settlements, Structures, Form and Layout, in *The Anglo-Saxons from the Migration Period to the Eighth Century: an ethnographic perspective*, ed. J. Hines (Woodbridge: Boydell, 1997), 101–124 (113–114).

[25] M. Clegg Hyer and G.R. Owen-Crocker, 'Making and Using Texiles', in *The Material Culture of Daily Living in the Anglo-Saxon World*, eds M. Clegg Hyer and G.R. Owen-Crocker (Exeter: University of Press, 2011), 157–184.

[26] P. Walton Rogers, '6.3 The importance and organisation of textile production', in *Rural Settlement, Lifestyles and Social Change in the Later First Millennium AD: Anglo-Saxon Flixborough in its wider context*, Excavations at Flixborough vol. 4, ed. C. Loveluck (Oxford: Oxbow Books, 2007), 106–111; P. Walton Rogers, 'Textile Production', in *Life and Economy at Early Medieval Flixborough c. AD 600-100: the artefact evidence*, Excavations at Flixborough vol. 2, eds D.H. Evans and C. Lovekuck (Oxford: Oxbow Books, 2009), 281–316 (300); P. Walton Rogers, 'Textile Production and Treatment', in *Staunch Meadow, Brandon, Suffolk: a high status Middle Saxon settlement on the fen edge*, East Anglian Archaeology 151, ed. K. Wade (Bury St Edmunds: Suffolk County Council Archaeology Service, 2014), 285–294 (285, 294).

the furnishings of the church', or, '… she mends, cleans, and adds to Worcester's ecclesiastical vestments'.[27] The word translated as vestments or furnishings is *indumentum*. The earliest use of *indumentum* in English sources is found in Bede and Aldhelm, who both refer to vestments or clothing, but from the 10th century the term also referred to ecclesiastical furnishings used as covers, so either translation is acceptable.[28] The renewing of old textiles referred to here is most likely to have meant replacing worn or damaged sections, or re-embroidering threadbare portions of ecclesiastical fabrics. It could also have indicated the complete replacement or recycling of textiles deemed unusable in their present state, as well as the making of new ones.

Christine Fell argued that the work required of Eanswitha was far too great for one woman so it must have been the whole household that did this work in lieu of rent.[29] However, if the estate was a central location with a variety of trades, it may have included an embroidery workshop and/or embroidery school that Eanswitha ran on behalf of the Church. If this was the case, Eanswitha would surely have had workers who carried out the renewing of textiles in the workshop, and others who ran the rest of the estate while she had overall responsibility for the workshop, the embroidery school (if it existed) and the estate.[30]

With the gradual development of urban areas, particularly from the 9th century, and the zoning of craft industries in places such as York,[31] embroidery-workers may have become part of a more specialised group that lived and worked on specific

---

27 'Ideo ego DENEBERHT episcopus, simul cum mea familia on Uuigerna ceastre, hoc decretum definiendo confirmavi, ut terram illam HEREFORDA, cujus quantitas est ii cassatorum, dabo EANSvVITHE possidendam quamdiu ipsa vivat, si illam post me superstitem in saeculo derelinquo, ea condicione ut ipsa sit semper subdita Uuigornensi aecclesiae, et ipsius familiae, hoc modo, ut semper illius aecclesiae indumentum innovet et mundet, et augeat, et postquam ipsa viam patrum incedat' terra supra dicta ad Uuigornensem aecclesiam absque omni contradictione conscripta est', W. de Gray Birch, ed., *Cartularium Saxonicum: a collection of charters relating to Anglo-Saxon history* vol. 1 (London: Chas. J. Clark, 1893), 307; A.J. Robertson, *Anglo-Saxon Charters*, 2nd edn (Cambridge: Cambridge University Press, 1956; repr. 2009), 378; C. Fell, *Women in Anglo-Saxon England*, (London: British Museum Publications, 1984), 42. The alternate translations are from Robertson and Fell, respectively.
28 [Rev.] J.A. Giles, *The Complete Works of Venerable Bede, In the Original Latin, Accompanied by a new English Translation: vol. IX Commentaries on the Scriptures* (London: Whittaker & co., 1844), 50, 178, 181; S. Gwara and R. Ehwald, eds, *Corpus Christianorum Series Latina: Aldhelmi Malmesbiriensis prosa de Virginitate: cum glosa latina atque anglosaxonica*, cxxiv A (2001), chapter LV, 715; R.E. Latham, D.R. Howlett, and R.K. Ashdowne, eds, *Dictionary of Medieval Latin from British Sources* (Oxford: Oxford University Press, 1997–2013), fascicule V I–L, 1340.
29 Fell, *Women in Anglo-Saxon England*, 42.
30 For references to textile workers on estates see: D. Whitelock, trans. and ed., *Anglo-Saxon Wills* (Cambridge: Cambridge University Press, 1930), 77; D. Herlihy, *Opera Muliebria: women and work in Medieval Europe* (London: McGraw-Hill, 1990), 77–83; P. Stafford, *Queen Emma and Queen Edith: queenship and women's power in eleventh-century England*, 2nd edn (Oxford: Wiley Blackwell, 2001), 109.
31 C. Dyer, *Making a Living in the Middle Ages: the people of Britain 850-1520* (London: Yale University Press, 2009), 65; Walton Rogers, *Textile Production at 16-22 Coppergate*, 1821–1829.

streets. The focus would most likely have been the production of embroidery to commission or for trade. However, at Six Dials, Southampton, the evidence points to craft industries being organised by household rather than zone.[32] Here and at other urban centres, women (and it is worth pointing out here that there is no evidence for male embroiderers in the British Isles before the Conquest) would have lived throughout the town and embroidered in the home or a workshop at the front of the property. Embroidery work may have been the main source of income for the household or it may have supplemented other forms of work. If these were small independent workshops they would have lacked storage for raw materials and finished products. This leads to the supposition that the patron would supply materials for work to be completed on a commission basis a job at a time, analogous to when Edward the Confessor (d. 1066) commissioned a goldsmith and bishop, Spearhafoc, to make him a crown. We are told that the king gave Spearhafoc the materials he needed to make the crown and he promptly ran off with them.[33] Materials could be stored only in larger organised workshops or at the residences of a patron; indeed, the patron might decide to gather workers together at their own property for larger projects. Again goldsmithing provides an analogy: we know that King Edgar (d. 975) employed and housed (possibly at King's Sombourne), skilled smiths to make a reliquary for Winchester.[34]

Convents also seem to have played a role. The evidence does not indicate commercial production but a form of training centre, or 'boarding school' to use Stephanie Hollis's term, where aristocratic and royal women were taught to embroider.[35] On the north side of the church at Whitby a number of buildings have been excavated that Sarah Foot suggested could have been used by women for domestic activities including weaving and sewing.[36] If she is right, it is probable that they would also have been used for embroidery work. Foot also discusses the role of the *nunna*, a somewhat obscure

---

32 C. Scull, 'Urban Centres in Pre-Viking England?', in *The Anglo-Saxons from the Migration Period to the Eighth Century: an ethnographic perspective*, ed. J. Hines (Woodbridge: Boydell, 1997), 269–310 (274, 277); Hinton, *Southampton Finds volume 2*, 98; A.D. Morton, *Excavations at Hamwic: Vol. 1 Excavations 1946–83, excluding Six Dials and Melbourne Street*, CBA Research Report 84 (London: Council for British Archaeology, 1992), 57.

33 D.A. Hinton, *Gold and Gilt, Pots and Pins: possessions and people in Medieval Britain* (Oxford: Oxford University Press, 2005), 142; J. Stevenson, ed., *Chronicon monasterii de Abingdon*, 2 vols, Rolls Series (London: Longman, Brown, Green, Longmans and Roberts, 1858), 1, 462–463.

34 A. Campbell ed., *Frithegodi monachi Breuiloquium vitae Beati Wilfredi = et Wulfstani cantoris Narratio metrica de Sancto Swithuno* (Turici: In Aedibus Thesauri Mundi, 1950), 141 (lines 7–11); F. Barlow, M. Biddle, O. von Feilitzen and D.J. Keene, *Winchester in the Early Middle Ages: an edition and discussion of the Winton Domesday* (Oxford: Oxford University Press, 1976), 466.

35 S. Hollis, 'Wilton as a Centre of Learning', in *Writing the Wilton Women: Goscelin's Legend of Edith and Liber confortatorius*, ed. S. Hollis with W.R. Barnes, R. Hayward, K. Loncar and M. Wright (Turnhout: Brepols, 2004), 307–338 (332).

36 S. Foot, *Monastic Life in Anglo-Saxon England c. 600–900* (Cambridge: Cambridge University Press, 2006), 118.

figure due both to the ambiguity of the rules governing women living religious lives outside the monastic enclave and later translations of the word itself. However, the general consensus is that a *nunna* was a woman who pursued a religious vocation outside the convent, either as an individual or part of a small group.[37] Documentary sources such as the *Liber Elienesis* show that these women embroidered as part of their vocation, working in their dwellings which may have extended to an estate (see the discussion of Æthelswith below).

One example of such a woman is Æthelswith. Her story appears in the *Liber Eliensis*, a 12th-century document that records the history of the community and church at Ely through the eyes of one of its monks. The monk collated pre- and post-Conquest documents from the monastery's library to form the work. The entry in question is from the 11th-century will of Leofflaed, the daughter of Ealdorman Byrhtnoth, and wife of Oswi, which was copied out in the *Liber Eliensis*. Leofflaed willed that upon her death, and if her daughters did not marry, certain land should be given to the church at Ely. One daughter did marry, but the other, Æthelswith, did not, taking lay orders instead. As a result, her portion of land, Stechworth, went to the Church and in return the monks gave her Coveney, a berewick (outlying estate) of Ely. Whilst there, Æthelswith spent her time in 'great seclusion' and produced gold embroidery and weaving with the help of young women, making, amongst other things, a chasuble of white fabric which she paid for at her own expense.[38]

Æthelswith is presented as working with young women (Latin: *puellulis*, a diminutive of *puellae*, girls) which may indicate that she was running a school or training workshop.[39] This is the only direct reference in the literature to suggest the establishment of such an institution. Æthelswith was probably a *nunna* who lived a religious life with a small group outside the monastic enclave. That she paid for at least one piece of embroidery out of her own wealth is also interesting. It indicates that other pieces may have been commissioned and paid for by other wealthy patrons, plausibly including the Church. If this is true, Æthelswith would have needed others to help her complete the commissions just as Eanswitha did. Therefore, it would be logical for Æthelswith to establish a form of school similar to the convent educational system where perhaps she had been trained. Indeed, it may be that Æthelswith was following an established tradition for high-ranking women. We do not know what rank the young ladies in her care were, but it would seem likely that they were of a similar social status to Æthelswith herself. It could be that the parents of those being

---

37 S. Foot, *Veiled Women vol 1: the disappearance of nuns from Anglo-Saxon England*, 2 vols. (Aldershot: Ashgate, 2000), 104–10.

38 '*Cui tradita est Coveneia, locus monasterio vicinus, ubi aurifrixorie et texturis secretius cum puellulis vacabat, que de proprio sumptu albam casulam suis manibus ipsa talis ingenii peritissima fecit*', E.O. Blake, ed., *Liber Eliensis* (London: Royal Historical Society, 1962), 158; J. Fairweather, trans., *Liber Eliensis: a history of the Isle of Ely from the seventh century to the twelfth* (Woodbridge: Boydell Press, 2005), 187–188.

39 Fairweather, *Liber Eliensis*, 188.

educated at Coveney paid towards living costs, and that this also contributed towards enabling Æthelswith to buy materials and other goods.

Archaeological evidence for early medieval Ireland indicates that textile production, and possibly embroidery, was set up rather differently than in Anglo-Saxon England. A detailed survey of early medieval dwellings and settlements in Ireland suggests that social status, gender and a strict organisation of social space were inextricably linked.[40] Areas appear to have been zoned, particularly within enclosure sites where set areas were designated to particular activities. An 8th-century round-house in Moynagh Lough, Co. Meath, where specific implements and artefacts were found concentrated in particular locations provides an example. The distribution of domestic craft implements indicated that these activities took place only in the southern, sunnier, half of the building. Moreover the distribution of textile-related finds indicated that the most likely place for textile associated activities was immediately adjacent to domestic buildings.

The authors of this survey argue that society was strictly hierarchical and as such, the types of craft work that each settlement was allowed to engage in was controlled by the social status of its inhabitants. In this context it would have been likely that certain settlements were allowed to produce textiles, and, we might infer, possibly also decorate them with embroidery, which would then have been distributed through exchange or trade to settlements that were not legally able to make them.

The Irish study also suggested that those Scandinavian towns that developed in Ireland during the later part of the period (900–1100), were still highly organised, with specific streets designated to particular craft activities.[41] Thus during the later period, areas associated with the new Scandinavian elites began to resemble zoned urban settings in Anglo-Saxon England like York, many of which were also run by Scandinavian incomers. On the other hand, evidence for more rural parts of Ireland suggests that its highly structured, hierarchical system was limiting by comparison to England's settlement development, which focused to a greater extent on prioritising trade relations and productivity. In both countries the specific circumstances in which production took place must have impacted on the evolution of embroiderers' working environments, leading cottage industry and workshop settings to develop along particular lines.

The embroidery-workers

All documentary sources point to women being the sole producers of embroidery in England and Ireland during the early medieval period. In Ireland for example, the law

---

40 A. O'Sullivan, F. McCormick, T.R. Kerr, L. Harney and J. Kinsella, *Early Medieval Dwellings and Settlements in Ireland, AD 400-1100*, British Archaeological Reports, International Series 2604 (Oxford: Archaeopress, 2014), 9.
41 O'Sullivan *et al.*, *Early Medieval Dwellings*, 13, 28, 32, 33, 64–65, 104, 109.

code *Corpus Iuris Hibernici* stated that, 'the lawful pledge-interests of an embroidery-needle among the Féni: for ornamental work, there is paid up to the value of an ounce of silver, for every woman who is an embroideress deserves more profit than even queens'.[42] This is corroborated by the archaeological evidence where textile and possible embroidery equipment is found only in female adult and child burials. Embroidery-workers can be split into two groups: those classed as professional embroiderers who produced embroidery as a form of living; and those of elite and royal circles, some of whom were later recognised as saints, whose facility at producing embroidery is referred to as a worthy accomplishment. In all documented cases embroiderers were held in high esteem: unlike producers of textiles, there is no evidence that embroidery-workers were ever classed as slaves.

As already discussed, the earliest evidence for a professional embroiderer is the 9th-century charter naming Eanswitha. The next is Æthelswith, who appears in an 11th-century will recorded in the 12th-century *Liber Eliensis*. Following on from that are the oft-cited Domesday Book references to Ælfgyth and Leofgyth. Domesday testifies that Leofgyth 'made and makes' gold-work for the King and Queen, '*Leuiede fecit 7 facit aurifrifiŭ regis 7 reginæ*'.[43] It may be that Leofgyth was one of the embroidery-workers who embellished items such as the many royal garments attributed to Queen Edith in the *Vita Edwardi*,[44] and then continued to work for King William and Queen Matilda after the Conquest.

In her will Queen Matilda (1031–1083) mentioned a textile worker, who may also have been an embroidery-worker.[45] The queen had commissioned a chasuble from a woman identified as Alderet's wife. Unfortunately, no details of the processes are specified so we do not know if this involved embroidery; nor is Alderet's wife's name given.[46] It is known that they lived in Winchester, but nothing beyond that.

The *Liber Eliensis*, the basis of what we know of the embroiderer Æthelswith, is a rich source of untapped information regarding embroiderers. In an inventory from January 1134, which listed everything the church at Ely owned at that time (some of which could thus have been produced and donated earlier), there are two further mentions of embroidery-workers with links to the monastery at Ely. The first is Liveva,

---

42 *Techta fuillema gill snaite druinige la Feine: imdenmaib direnar corruice log nuinge argit, air is mo do thorbu dosli cach ben bes druinech olldaite cid rigna*, D.A. Binchy, ed., *Corpus Iuris Hibernici: ad fidem codicum manuscriptorum recognovit*, volume 2 (Dublin: Dublin Institute for Advanced Studies, 1978), 464.2–3; trans.: M. Zumbuhl, *Lexis of Cloth and Clothing Project*, Unpublished conference paper (2011).
43 Thorn and Thorn, '6: Wiltshire', 74b.
44 '*Obsecuta est illi tamquam filia regina egregia eumque a principio sue desponsionis diuersis in opere redimiuit*ᵃ *ornamentis*', F. Barlow, trans. and ed., *The Life of King Edward, who rests at Westminster*, 2nd edn (London: Nelson and Sons, 1962), 15.
45 L. Musset, *Les Actes de Guillaume*, 112–113.
46 Neither his nor her name are recorded in the database *Prosopography of Anglo-Saxon England* (2010) http://www.pase.ac.uk/index.html (Accessed: 23 December 2011).

who is given the title of *aurifrixatricis*, translated by Fairweather as gold-worker.[47] She owned and gave to the church two albs (white vestments worn by clergy) decorated with *pallio* (singular *pallium*). The word *pallium* is a little hard to pin down, and is perhaps best understood as a fine, decorated ecclesiastical textile.[48]

Listed in the same inventory are two woollen *dorsalia* (dorsals): probably meaning precious cloths hung behind or above an altar.[49] The entry in the *Liber Eliensis* describes the hangings simply as 'woollen', so we do not know whether they were woven from wool, or embroidered in wool. They are described as originally belonging to Ingrith, an embroidery-worker with experience in gold-work.[50] If the hangings were woven it is likely that Ingrith did not produce them herself but had them made to commission or even inherited them. Since we do not know what kind of household Ingrith ran we do not know whether she had other women to make the hangings for her. If, however, they were embroidered (rather than woven) in wool she is perhaps more likely to have stitched them herself, although since she was an experienced gold-worker producing intricate gold embroidery, she may not have had time to produce the woollen hangings as well (see pp. 120–121). Whichever way the hangings came to be in Ingrith's possession, she probably used them first in her own home and then transferred them to the church at an agreed point, perhaps in a similar manner to the better known Byrhtnoth wall-hanging (either woven or embroidered), which was given to the community at Ely by Byrhtnoth's wife Æthelflæd after his death at the battle of Maldon against the Vikings in 991.[51]

The inventory also notes that by 1134 the monks at Ely had bought an *infula* made of an expensive silk called *purpura*. Fairweather translates *infula* as chasuble but such a meaning seems to be a late and infrequent use. It is more likely a mitre or the decorated band attached to a mitre. C.R. Dodwell surmised that *purpura* was a form of shot-silk taffeta (not necessarily purple in colour). The *Liber Eliensis*

---

47 Fairweather, *Liber Eliensis*, 358. Also see Latham *et al.*, *Dictionary of Medieval Latin from British Sources*, fascicule I A–B, 164.
48 The Lexis of Cloth and Clothing Project (2014) http://lexissearch.arts.manchester.ac.uk/entry.aspx?id=3575 (Accessed: 16 January 2015) defines a pallium as a) a narrow ecclesiastical vestment, the insignia of the pope of archbishops, b) an ecclesiastical cloak or mantle, or c) a form of fine material that might, for example, be used to cover a chalice; which we might here understand to have been applied to the two albs. Fairweather translates simply as 'pall-cloth': *Liber Eliensis*, 358; and see '*Et lxiiii albe parate pallio ... ii Liveve aurifrixatricis*', Blake, *Liber Eliensis*, 293. See also Latham *et al.*, *Dictionary of Medieval Latin from British Sources*, fascicule IX P-Pel, 2088.
49 Or alternatively, but less likely, hangings for the backs of chairs, see A.J. Schulte, 'Altar Screen', in *The Catholic Encyclopedia* (1907) http://www.newadvent.org/cathen/01356d.htm (Accessed: 22 November 2018). The entry states that such cloths could be embroidered, cloth of gold or tapestries.
50 '*Et ibi xxxiiii bona dorsalia de lana, ... ii Ingrithe aurifris*', Blake, *Liber Eliensis*, 294; Fairweather, *Liber Eliensis*, 359.
51 The hanging is originally mentioned in an inventory in the *Liber Eliensis*: Blake, *Liber Eliensis*, 136, Fariweather, *Liber Eliensis*, 163. For discussion of the donation and possible movement of the Brythnoth wall-hanging to Ely Cathedral see M. Budny, 'The Byrhtnoth Tapestry or Embroidery', in *The Battle of Maldon AD 991*, ed. D. Scragg (Oxford: Blackwell, 1991), 263–278.

suggests this piece had been well-embroidered, '... i infula de purpura bene brusdata, quam fratres ecclesie emerunt ab Ælboldo, burgensi Theodfordie'.[52] If the *infula* was an embroidered band of silk it would have been bought in order to attach to the mitre and the maker may also have been a producer of secular pieces, such as the gold woven strips decorated with gems that Byrhtnoth detached from his cloak and donated to Ely in c. 991, and the gold embroidered bands that King Edgar detached from his boots and also donated to Ely.[53] It is this sort of embroidery that Liveva might have produced. In this case, however, we do not know the name of the worker. We do know that the monks bought the *infula* from a man named Ælfbold, a burgess (full citizen) of Thetford, a berewick or outlying estate of Ely. It is likely that Ælfbold was either a merchant or craftsman, and he may have been a middleman for the sale of religious or secular textiles. He may have produced the *infula* himself, but this is unlikely in the light of the lack of evidence elsewhere for male embroiderers. It is more likely that he ran an embroidery workshop or controlled a smaller dispersed work-force. Staniland hypothesised that embroidery workshops had been established in London from early in the medieval period, but there is no clear evidence before the 13th century.[54]

Additional to evidence of professional embroiderers, or their work, we hear about elite and royal women embroiderers. Such women were trained in the art of embroidery from childhood. Sheriff Godric's daughter discussed above provides an example. Saints' lives and royal biographies provide anecdotes of those who were supposed to have great skill in the art. Some of this information is likely to have been exaggerated in order to enhance the character of the central figure, and it is clear that we need to approach these sources with an awareness of their different objectives. However, the presentation of such women as embroiderers is credible. Embroidery was deemed a worthy and appropriate accomplishment for women of status. Æthelswith was not the only high ranking woman to give up her life in society and take religious or lay orders. An earlier example in the *Liber Eliensis* is Queen (later Saint) Æthelthryth (d. 679), who, 'being skilled in handiwork, she made with her own hands ... by the technique of gold embroidery, an outstanding and famous piece of work'.[55] Æthelthryth was the daughter of Anna, King of the East Angles and wife first of Prince Tondberht (d. c. 655) and then the 15-year-old Ecgfrith, King of Northumbria,

---

52 Blake, *Liber Eliensis*, 293; Fairweather, *Liber Eliensis*, 357; C.R. Dodwell, *Anglo-Saxon Art: a new perspective* (Manchester: Manchester University Press, 1982), 145–150.
53 'duabus laciniis pallii sui, pretioso opere auri et gemmarum contextis', Blake, *Liber Eliensis*, 135; Fairweather, *Liber Eliensis*, 162; 'et xi albe parate subterius aurifrixo leviter, i de caligis Ædgari regis', Blake, *Liber Eliensis*, 293; Fairweather, *Liber Eliensis*, 358.
54 Staniland, *Medieval Craftsmen*, 5, 10, 27–32, 49–53.
55 'Insuper opus eximium atque preclarum, stolam videlicet et manipulum similis materie ex auro et lapidibus pretiosis propriis, ut fertur, manibus docta auritexture ingenio fecit eique ob interne dilectionis intuitum pro benedictione offerendum destinavit ...', Blake, *Liber Eliensis*, 24; Fairweather, *Liber Eliensis*, 30.

whom she married in 660 before leaving him to become a nun at Coldingham.[56] In 673 she founded the monastery at Ely. That Æthelswith and Æthelthryth were praised for embroidering whilst living a religious life demonstrates that this activity was prized and associated with piety.

The convent of Wilton, which was founded in AD 890 by King Alfred (d. 900), educated women of high status in the skill of embroidery. It survived the 10th-century Benedictine reforms, and continued to be a place of worship and education through to at least the 11th to 12th centuries when Edith (also known as Matilda), and Gunnilda, daughters of Saint Margaret (c. 1045–1093) and Malcolm III of Scotland (d. 1093), were educated there, as their mother had been.[57] Margaret was the granddaughter of King Ethelred (d. 1016) and therefore of royal Anglo-Saxon lineage. She spent the latter part of her childhood at the court of King Edward the Confessor (d. 1066) and Queen Edith (d. 1075). In 1066 she fled to Scotland and married into the Scottish royal family. Margaret's biographer, Theodoric, who was also her confessor, wrote, 'her chamber was like a workshop of a heavenly artist; there were copes for singers, chasubles, stoles, altar-cloths and other priestly vestments and church ornaments were always to be seen, some in course of preparation, others, worthy of admiration, already completed'.[58] Looking past the laudatory tone, the volume of work described suggests that Margaret embroidered with a group of women.

Earlier, William of Malmesbury, said of Edward the Elder's (d. 924) nine daughters, 'All the daughters had been brought up to devote most of their time in their childhood to letters, and thereafter to acquire further skill with distaff and needle, that with the support of these arts they might pass their girlhood in chastity'.[59] The first daughter of Edward's second marriage, Eadflæd, took the veil and the third of the same marriage, Æthelhild, took lay orders. Both were buried at Wilton next to their mother Ælfflæd. This is the same Ælfflæd who commissioned

---

56 See R.C. Love, 'Æthelthryth', in *The Wiley Blackwell Encyclopaedia of Anglo-Saxon England*, 2nd edn, eds M. Lapidge, J. Blair, S. Keynes and D. Scragg (Oxford: Wiley Blackwell, 2014), 19–20.

57 D.K. Coldicott, *Hampshire Nunneries* (Chichester: Phillimore, 1989), 2, 4; H. Leyser, *Medieval Women: a social history of women in England 450-1500* (London: Weidenfeld & Nicolson, 1995; repr. London: Phoenix, 1997), 82; B. Yorke, *Nunneries and the Anglo-Saxon Royal Houses* (London: Continuum, 2003), 152.

58 'Hiis rebus, id est quae ad divinae servitutis cultum pertinebant, numquam vacua erat illius camera; quae, ut ita dicam, quaedam caelestis artificii videbatur esse officinal. Ibi cappae cantorum, casulae, stolae, altaris pallia, alia quoque vestimenta sacerdotalia, & ecclesiae semper videbantur ornamenta', in *Vita Auct. Theodorico Monacho Dunelm. Confessario ipsius Sanctae, ad filiam Mathildem Angliae Reginam, ex Membraneo nostro Ms. Valcellensii*, Acta Sanctorum Database (2012) http://acta.chadwyck.co.uk (Accessed: 6 August 2012).

59 'Filias suas ita instituerat ut litteris omnes in infantia maxime uacarent, mox etiam colum et acum exercere consuescerent, ut his artibus pudice impubem uirginitatem transigerent', R.A.B. Mynors, R.M. Thomson and M. Winterbottom, eds, *William of Malmesbury, Gesta Regum Anglorum: The History of the English Kings* (Oxford: Clarendon Press, 1998), 198–201.

what are now known as the Cuthbert embroideries. Ælfflæd had close links with Wilton. She had been brought up there and returned to live there for a year when Edward banished her from court before AD 920.[60]

Wilton had close links to other royal women. In the mid- to late 10th century Edith (c. 961–984) the daughter of King Edgar (d. 975), was educated there.[61] She was a nun whilst her mother was abbess, later acceded to the role herself, and became Wilton's patron saint after death. Her embroidery skills are outlined in the biography *La Légende de ste. Édith*, which describes her many accomplishments, including a reference to her embroidering with gold thread and precious stones.[62] It also includes a description of a particular piece:

> Saint Edith embroidered an alb with gold-work, jewels and pearls, some of which were small English pearls. The hem of the alb was worked with golden figures of the apostles standing around the Lord, who was seated in their midst, while Edith prostrated herself in the role of the suppliant Mary, kissing the feet of the Lord.[63]

The picture painted here reminds one of the 11th-century title page illustrations of books where the author or donor is placed at the feet of Our Lord or a saint; for instance, a manuscript from the Cotton British Library MS Titus D. XXVI, fol. 19v., where the author, Ælfwine, a monk of Winchester, is seated at the feet of Saint Peter.[64]

At the end of the Anglo-Saxon period a second Edith joined the royal family by marriage to Edward the Confessor (d. 1066). Supposedly named after Edith of Wilton, she was daughter of the powerful Earl Godwin of Kent (d. 1053) and was royally connected through her Scandinavian mother. She was also reputed to have great skill in working embroidery, '… *Edeiha diligitur, que uersu et prosa celebres et eximia et opera et picture altera erat Minerua*'.[65] Although the text is ambiguous, Dodwell has argued that because of the reference to Minerva, who was known in the ancient world for her skill in textile production, *opus* can be read as needlework and *pictura*

---

60 E. Coatsworth, 'The Embroideries from the Tomb of St Cuthbert', in *Edward the Elder 899–924*, eds N.J. Higham and D. Hill (London: Routledge, 2001), 292–306 (296); S. Sharp, 'The West Saxon tradition of dynastic marriage: with special reference to Edward the Elder', in *Edward the Elder*, eds Higham and Hill, 79–88 (82).
61 Coatsworth, 'Embroideries from Tomb of Cuthbert', 396; Barlow, *Life of King Edward*, 23.
62 '… *uox cignea, canor angelicus, mellita facundia, generosum et ad omnia capax ingenium, legendi intellectuosa flagrantia; manus pingendi, scriptitandi, dictitandi tam decentes, quam artificiose; digiti aurifices, gemmarii, citharedi, citharizantes…*', Goscelin, 'La Légende de S. Édith', *Analecta Bollandiana*, vol. 56 (Bruxelles: Société des Bollandistes, 1938), 68–69; Dodwell, *Anglo-Saxon Art*, 53, 55; Stafford, *Queen Emma and Queen Edith*, 257–258.
63 '*Inter quae fecerat ibi ex bisso candidissimo albam … praestantissimam auro, gemmis, margaritis ac perulis Angligenis a summo contextam … circa pedes aureas apostolorum ymagines Dominum circumstantes, Dominum medium assidentem, se uice suplicis Mariae affusam, dominica uestigia exosculantem*', Goscelin, *La Légende de S. Édith*, 79; Dodwell, *Anglo-Saxon Art*, 57.
64 Dodwell, *Anglo-Saxon Art*, 57, 59.
65 Barlow, *Life of King Edward*, 14.

as embroidered pictures. In the biography of her husband, commissioned by Edith herself, she is described as embellishing, or commissioning embellishment of, items such as the robes that her husband wore.[66] Disregarding the elaborate nature of the writing, which is similar in vein to the biography of Saint Margaret, the text suggests Edith could embroider and probably did so with the help of others. This is supported by the fact that she was educated at Wilton.

Wilton was as one of a group of convents with royal connections clustered around Winchester.[67] Stephanie Hollis has suggested that other convents patronised by the Wessex royal family, including Barking and Horton, and Shaftesbury and Nunnaminster, had monastic schools.[68] Other houses such as Romsey, which had royal abbesses and nuns, may plausibly have taught embroidery, but the documentary evidence does not survive.

We also know about Æthelwynn, a married woman who wrote to Saint Dunstan (c. 924–988) asking him to design an embroidery pattern for a priest's stole that she and her girls (*puellae*) could embroider in gold. Could this be another indication of a school or workshop run by a high status woman, as suggested for Æthelswith at Coveney? There are two sources, one in *Memorials of Saint Dunstan* and the second in *William of Malmesbury's Saints' Lives*,[69] and each is making different points using the same story. The latter gives the clearer and more succinct account of the incident, 'Dunstan was requested by a married lady called Æthelwynn to come to her house to draw a picture for a priest's stole, so that her girls could use it as a pattern in sewing the gold design. The Latins call it "feather-work"'.[70] It would seem that Æthelwynn and her girls were embroidering these items not just as an appropriate form of elite pastime but to donate to an ecclesiastical centre or even a specific priest, perhaps treating it as a form of vocation. The description of the commissioned design with its pictures of saints calls to mind the surviving fragments from Worcester which have been dated to the 12th century but are worked in a hybrid late Anglo-Saxon early *Opus Anglicanum* style, and the stole that King Athelstan is thought to have donated to Saint Cuthbert.[71] The use of the term *opus plumarium* is of interest because the embroideries Æthelwynn and her girls were creating demonstrates the development

---

66 Dodwell, *Anglo-Saxon Art*, 258; Barlow, *Life of King Edward*, 15.
67 Stafford, *Queen Emma and Queen Edith*, 257, 258.
68 S. Hollis, 'Wilton as a Centre of Learning', 336.
69 W. Stubbs, ed., *Memorials of Saint Dunstan Archbishop of Canterbury* (London: Longman, 1874), 20–21; M. Winterbottom and R.M. Thomson, eds, *William of Malmesbury Saints' Lives: Lives of SS. Wulfstan, Dunstan, Patrick, Benignus and Indract* (Oxford: Clarendon Press, 2002), 182–183.
70 'Rogatus est Dunstanus a quadam matron Ethelwinna nomine domum suam uenire, quatinus in casula sacerdotali faceret picturam, unde puellae suae insuendi auri traherent formam. Opus plumarium uocant Latini', Winterbottom and Thomson, *William of Malmesbury Saints' Lives*, 182, 183.
71 For images of the stole found at Worcester see M. Grundy, 'Needlework Stitched nearly 1,000 Years Ago is Brought Back to Life', *Worcester News* (2008) http://www.worcesternews.co.uk/features/3562791.print/ (Accessed: 24 April 2010).

of contouring through shading. As Elizabeth Coatsworth has persuasively argued, *opus plumarium* was probably a term used to describe areas of embroidery infilled with different shades of coloured thread to create the impression of muscles or varied coloured effects.[72] The development of this approach, which is used to great effect in the later *Opus Anglicanum*, can also be seen in early stages of development on the Worcester and Cuthbert stoles, both of which use gold and silk threads and contain figures whose faces are defined through the use of different colours.

Finally, Christie referred to an Ælgitha of Northampton, usually called Ælfgiva, who was either the mistress or first wife of King Cnut and is supposed to have embroidered a number of altar-cloths which Cnut and Emma, his later wife, gave to the monastery at Croyland.[73] However, she does not give a primary source. The only early source, 'Ingulphs' chronicle of the abbey of Croyland with the continuations by Peter of Blois and anonymous writers', has been identified as partially a 14th-century forgery.[74] It is thus difficult to confirm Ælgitha's skills, but as the daughter of ealdorman Ælfhelm (a high status noble),[75] she is likely to have been taught to embroider from an early age and it is not unlikely that she would have embroidered and commissioned work for ecclesiastical settings.

There has been some debate as to whether elite named women were really such prolific embroiderers, or whether attendant ladies or professionals such as Leofgyth were largely responsible. It would be logical for others to work on projects with or for the lady in question. If a lady was truly as industrious as some are made out to be, for example Queen Edith who is said to have embroidered all of her husband's clothing, there would have been little time left for anything else![76] To put this into context, in the mid-1980s Helen M. Stevens, a professional embroiderer, recreated one of the repeating units of the arcade from the Maaseik embroideries. The piece consists of couched gold-work, split stitch and stem stitch worked in silk thread. At the end of the eighteen-month project Stevens concluded that if the embroideries were worked in a professional manner, one such unit could be completed in one week with between 4.5 and 5 hours' close work each day. Stevens suggested this would be the maximum amount of time an embroiderer could work and still maintain a consistently high, professional standard of embroidery.[77] From this evidence it can

---

72 E. Coatsworth, 'Opus What? The Textual History of Medieval Embroidery Terms and Their Relationship to the Surviving Embroideries', in *Textiles, Text, Intertext: essays in honour of Gale R. Owen-Crocker*, eds M. Clegg Hyer and J. Frederick (Woodbridge: Boydell Press, 2016), 43–67 (54–59).
73 Christie, *English Medieval Embroidery*, 32.
74 H.T. Riley, trans., *Ingulph's chronicle of the abbey of Croyland: with the continuations by Peter of Blois and anonymous writers* (London: H.G. Bohn, 1854).
75 S. Keynes, 'Cnut', in *Wiley Blackwell Encyclopedia of Anglo-Saxon England*, eds Lapidge et al., 111–112.
76 Barlow, *Life of King Edward*, 15.
77 H.M. Stevens, 'Maaseik Reconstructed: a practical investigation and interpretation of 8th-century embroidery techniques', in *Textiles in Northern Archaeology: NESAT III*, eds P. Walton and J.P. Wild (London: Archetype Publications, 1990), 57–60.

be proposed that this piece, one of four incomplete strips, would have taken over eight weeks to complete. Even the simpler pieces of embroidery would have taken substantial amounts of time to create, as archaeologist Chrystel Brandenburgh has demonstrated. As part of a research project, Brandenburgh recreated a headdress she discovered in the stores of the National Museum of Antiquities in Leiden, the Netherlands. The hat was discovered at Dokkum-Berg Sion in the Netherlands, and has been radiocarbon dated to between AD 600 and 900. Its seams were joined together and covered with two forms of looped stitch. It took Brandenburgh seven hours to sew the headdress. She points out that it would have taken less than half that time if a more basic stitch had been used.[78] This sort of experiment usefully demonstrates the intricacies of embroidery production. It is technically complex and in most cases, such as the production of ecclesiastical vestments (Durham C, D, Worcester) and the Maaseik bands, would have constituted full-time employment. The costs involved in training women in the required skills and technical knowledge and commissioning such pieces must have been high.

What the surviving embroideries tell us
Materials (and developing settlement patterns)

The different materials used in embroidery and the functional and symbolic use of embroidery within a variety of contexts in early medieval England have already been discussed. An analysis of materials can provide valuable insights into the working environments in which embroidery took place, as well as being relevant to our understandings of trade routes and the movement of people and ideas. The 7th-century piece from Sutton Hoo (Sutton Hoo A) is the earliest example from my corpus to yield information of this type. This is a wool embroidery worked on a wool ground fabric. The embroidery is a variation of looped stitch and has been used to cover and reinforce a seam join. The ground fabric is woven with a Z-spun warp thread and an S-spun weft thread (see Glossary). As no selvedge has survived it is not possible to distinguish which system is the warp and which is the weft.[79] The thread count varies between 17–19 threads per cm in one direction and 13–15 threads per cm in the other.[80] By taking a section of the ground fabric apart, Elisabeth Crowfoot was able to discern the textile's woven pattern as an irregular broken diamond twill (see Glossary). She also noted that the embroidery fibre was

---

78 C. Brandenburgh, 'Old Finds Rediscovered: two early medieval headdresses from the National Museum of Antiquities, Leiden, the Netherlands', *Medieval Dress and Textiles*, 8 (2012), 25–48 (25, 46).
79 E. Crowfoot, 'The Textiles', in *The Sutton Hoo Ship-Burial volume 3*, eds R. Bruce-Mitford and A.C. Evans (London: British Museum Publications, 1983), 404–479, (421).
80 Crowfoot, 'The Textiles' (1983). The published report does not indicate which thread count belongs to which system, Z-spun or S-spun.

a single thread of wool similar to that used for the ground fabric.[81] In 1965–1970 fibre samples were sent for analysis at the Animal Breeding Research Organisation at the Roslin Institute, University of Edinburgh.[82] The results showed that the wool fibres were fine generalised medium wools with occasional coarser fibre. It was suggested that the wool was taken from a 'primitive sheep type'.[83]

The fabric had been dyed (although the results of dye tests were inconclusive), and the embroidery thread may have been dyed the same colour. This helps to distinguish the type of settlement in which the fabric was made and dyed. The wool was probably taken from sheep reared locally, and prepared, woven and dyed within the same settlement. The fabric would then have been made into the final object and embroidered on site. All of this indicates a high status community with the resources to produce (or maybe specialised in producing) good quality woollen fabrics that could be dyed and turned into items that, like Sutton Hoo A, were used by and/or buried with the local elite. Archaeologically, the excavated site of Brandon is a good comparative example. Here the production and dyeing of textiles was integrated within the village system, with weaving taking place within at least one area located next to the dye-works (see above). Here, too, there are possible dwellings where makers and embroiderers could have worked. Indeed, in the case of Sutton Hoo A, it is probable that the maker of the textile was also its embroiderer. This deduction is based on the evidence that the stitching is worked in the similar thread as that from which the object was woven, and the stitching is both decorative and functional: its main role is to join the seam together but it has been carried out in decorative style. Walton Rogers has argued that Brandon was producing textiles for its own inhabitants,[84] and it is likely that production worked in a similar way for the Sutton Hoo A community.

If this is true, we have an integrated system of working within a localised area and settlement. As we know that the embroidery thread was probably the same as that of the textile we might also surmise that in the case of Sutton Hoo A: the item was probably a commission. The wool was most likely spun in a batch and set aside for this project. It was then dyed while it was thread before being woven into a textile, and extra thread was placed in the vat so that the sewing and embroidery threads would match the colour of the textile once it was made. Dyeing both textile wool and

---

81 Crowfoot, 'The Textiles' (1983), 421, 422. She took part of the sample apart for close study. The report does not indicate how the embroidery thread is similar to the fabric.
82 Crowfoot, 'The Textiles' (1983), 417; University of Edinburgh, *History of the Institute*, http://www.roslin.ed.ac.uk/about-roslin/history-of-the-institute/ (Accessed: 13 October 2013); My Society, *Wool Textile Industry Levy*, http://www.theyworkforyou.com/debates/?id=1950-10-24a.2733.0 (Accessed: 13 October 2013).
83 M.L. Ryder, 'Appendix 1: Wool from Anglo-Saxon Sites', in *Sutton Hoo Ship-Burial volume 3*, eds Bruce-Mitford and Evans, 463–464 (463).
84 Walton Rogers, 'Textile Production and Treatment', 294.

extra thread together ready for use in the construction of the final piece, meant they could be given to the maker for both production and embellishment of the object, and then passed on to the recipient.[85]

The 8th- to 9th-century embroideries from Maaseik (Maaseik A, B and C) and the stole, maniple and possible headdress ribbons or girdle from Durham (Durham C, D and E), which probably date to 909–916, yield useful data concerning how production systems developed. The ground fabric is of interest here. The evidence from both groups suggests the use of batches of materials which could have been stored and used as required. This indicates a workshop setting where materials were bought in larger quantities from a number of different sources, or possibly from a single trader who, himself, sourced from multiple producers.

Mildred Budny and Dominic Tweddle convincingly argue that the Maaseik embroideries were all worked in the same workshop, on the basis of stylistic markers and evidence of how the work was constructed.[86] The technical attributes of all the embroideries are very similar, which bears out this conclusion. It is therefore interesting that the ground fabric used for each piece is different, and the embroidery threads, although all silk, may have come from different batches. Maaseik A (the arcade strips), was worked on a tabby-weave (see Glossary) linen constructed of 23 × 26 Z-spun threads per centimetre. This is slightly finer than the ground fabric used for Maaseik B (the roundel strips), which is a tabby-weave linen made from 26 × 20 Z-spun threads per centimetre. The ground fabric used to create Maaseik C (the four monograms), comprises a tabby-weave linen of 24 × 20 Z-spun threads per centimetre in a weight in between that of A and B.[87]

The Durham embroideries show similar variation in ground fabrics. Durham C was worked on a silk fabric of 22 × 22 threads per centimetre while Durham E was worked on a silk fabric of 25 × 25 threads per centimetre.[88] During her examination of the stole (Durham D) Plenderleith noticed that the ground fabric of the prophet Zachariah was the same as that of the maniple (Durham E), which is finer than that used for the headdress/girdle (Durham C) and other parts of the stole. This led Plenderleith to suggest that the stole was worked in sections and joined together once complete.[89] In

---

85 The Roslin report and Crowfoot's work indicates that one set of the woven threads (either warp or weft) were dyed and the other set were undyed, although we do not know which was which as there is no selvedge for us to use as reference. This proves that Sutton Hoo A could only have been dyed *prior* to being been woven. See Crowfoot, 'The Textiles' (1983), 421, 422; M.C. Whiting, 'Appendix 2: The Dye Analysis', in *Sutton Hoo Ship-Burial volume 3*, eds Bruce-Mitford and Evans, 465.
86 M. Budny and D. Tweddle, 'The Early Medieval Textiles at Maaseik, Belgium', *The Antiquaries Journal*, 65 (1985), 353–389 (372).
87 M. Budny and D. Tweddle, 'The Maaseik Embroideries', *Anglo-Saxon England*, 13 (1984), 65–96 (76).
88 Plenderleith, 'The Stole and Maniples (a) the technique', 383, 390. The conversion of threads per inch to threads per centimetre has been done by the author.
89 Plenderleith, 'The Stole and Maniples (a) the technique', 383.

fact, Christie had previously put forward a similar thesis in *English Medieval Embroidery*, pointing out that there are distinct lines within the embroidery work at the bases of the mounds on which the figures stand.[90]

Again, this suggests different types of fabric were being stored in the same workshop and were used as needed, and indicates that fabric was being bought and stored in batches. The silk threads may have been bought in the same way. Maaseik A and B make use of threads of different twists and plies; some of the red and green threads are Z-twisted (see Glossary) and on occasion S-plied (see Glossary), while other colours – the yellow, beige and blue – are untwisted.[91] The silk-work on Durham D and E utilises two types of silk thread. A thin, non-twisted version in various colours has been used for split stitch fillings of foliage, garments, the rims of the nimbi, and the faces and hands of the figures. A thicker silk thread in dark brown and dark green, with a slight S-twist, was used to stitch the stem-stitch outlines.[92] It would seem that different types of thread were selected for specific purposes. The untwisted silk thread is easier to split and therefore most useful when working split stitch. The slight twist of the second option allows it to lie flat without splaying out, making it ideal for sewing stem stitch since it is easier to bring the needle up next to it in order to make the following stitch in the line. Either different types of silk thread may have been bought and stored in the workshops; or perhaps, untwisted threads were bought, possibly in bulk, and made up in different forms as needed.[93]

Use of silk thread indicates workshop owners interacting with trading contacts with overseas networks. At this time silk thread had to be imported from Byzantium, the Middle East and China: only these countries knew how to produce silk and could turn it into thread and cloth, and they tightly regulated all aspects of the production process.[94] Balls and skeins of silk thread were found during excavations at the Fishamble Street in Dublin,[95] established by the Vikings during the 9th century.[96]

---

90 Christie, *English Medieval Embroidery*, 49. She concluded that each figure had been stitched by a different worker on a small frame that could be turned to make reaching different parts of the design easier, and the pieces were joined together once finished.
91 Budny and Tweddle, 'The Maaseik Embroideries', 76.
92 Baldwin Brown and Christie, 'S. Cuthbert's Stole and Maniple', 5; Plenderleith, 'The Stole and Maniples (a) the technique', 377, 380.
93 An experimental project the author is undertaking which recreates part of the Cuthbert maniple involves using the same threads for the split and stem stitches, but doubling up and slightly plying those used for the stem stitch to give the thickness and slight ply required. All available pre-made threads were found to be too thick and too tightly twisted.
94 M. Vedeler, *Silk for the Vikings*, Ancient Textiles Series, 15 (Oxford: Oxbow Books, 2014), 83, 97–100, 102–104.
95 F. Pritchard, 'Evidence of Tablet-Weaving from Viking-Age Dublin', Unpublished conference paper (2014).
96 E. Purcell and J. Sheehan, 'Viking Dublin: enmities, alliances and the cold gleam of silver', in *Everyday Life in Viking-Age Towns: Social Approaches to Towns in England and Ireland, c. 800–1100*, eds D.M. Hadley and L. Ten Harkel (Oxford: Oxbow Books, 2013), 35–60 (41).

It is likely that Scandinavian traders were importing such items into early medieval Ireland and England.

Silk was definitely entering north-western Europe through the Russian river systems through which the Scandinavians had established trading links on the silk route, possibly stretching as far as China itself.[97] Using a combination of exchange and these trading networks, workshops in early medieval England would have been able to obtain silk thread. It is likely that such threads were bought in bulk and kept carefully for important commissions, since the supply of silk may not have been continuous.

The gold thread used to create the Maaseik embroideries is interesting because it is made with a central core of cattle-tail hair wrapped with a nearly pure gold sheet that was hammered out flat and cut into fine strips measuring approximately 0.5 mm wide.[98] While the measurement is consistent with other examples of early medieval Anglo-Saxon gold thread, the use of a cattle-tail hair for the core is not. Elizabeth Plenderleith has noted that from the 11th century, threads of silver-gilt metal could be wound round a core of hair but she does not give any examples.[99] Durham C, D and E exhibit the more usual early medieval gold thread with a core of silk. It is not known whether gold thread used in embroidery at this time was imported, produced in Anglo-Saxon England, or sourced in both ways. There does seem to be some evidence for a combined approach. Gold thread wound round a silk core was produced in the Byzantine Empire and could have been obtained for use in Anglo-Saxon England via traders. During the later part of the period, gold thread was also produced in Italy, but this exhibits a rectangular cross-section[100] and was thus produced using different techniques to the gold thread with a circular cross-section used in Anglo-Saxon England. During excavations in Southampton a reel of gold thread with a silk core was found in an early medieval cesspit, but it may originally have come from a grave that had been disturbed during the 8th and 9th centuries (see Fig. 18).[101] At this period Southampton was a busy harbour town, and the lost or accidentally misplaced reel of thread could have come into the town through trading routes.

---

97 J.D. Richards, *Viking Age England* (London: B. T. Batsford/English Heritage, 1991; repr. Stroud: The History Press, 2010), 159; E. Andersson Strand, 'Northerners: global travellers in the Viking Age', in *Global Textile Encounters*, eds M.-L. Nosch, Z. Feng and L. Varadarajan (Oxford: Oxbow Books, 2014), 75–80; Vedeler, *Silk for the Vikings*, 58–62.
98 Budny and Tweddle, 'The Maaseik Embroideries', 76.
99 Plenderleith, 'The Stole and Maniples (a) the technique', 378.
100 P.M. de Marchi and A.F. Palmieri-Marinoni, 'Longobard Brocaded Bands from the Seprio: Production, Movement and Status', Unpublished conference paper (2014).
101 Hinton, *Gold and Gilt*, 87; P. Walton Rogers, 'Gold Thread', in *The Origins of Mid-Saxon Southampton: Excavations at the Friends Provident St Mary's Stadium 1998-2000*, eds V. Birbeck, R.J.C. Smith, P. Andrews and N. Stoodley (Salisbury: Wessex Archaeology, 2005), 68–69.

There are hints that Anglo-Saxon goldsmiths may have made gold thread as a side-line. Elizabeth Coatsworth and Michael Pinder, and David Hinton have discussed the employment of goldsmiths. While they could be found working in their local settlements during the early part of the period, changes to social structure over time led to goldsmiths working in ecclesiastical establishments, as part of elite or royal households, as independent metalworkers within urban settings, or as itinerant workers who travelled from place to place.[102] Within estate and royal settings gold-workers are quite likely to have made gold thread for their mistresses' use. Those based in religious establishments may also have made gold thread for nuns and *nunna* (see pp. 111–112), while those who moved from place to place would have been able to sell their wares to anyone who could afford it, as would those who set up independently within urban centres.[103] It would seem likely, then, that gold thread was at first imported into Anglo-Saxon England, but once local goldsmiths learnt how to make it, it would also have been locally produced (although they would have had to source core silk thread).

By the end of the early medieval period gold thread was often made not of gold but of silver-gilt. It is this type of thread that was used to decorate Durham F and G, two fragments of embroidery found in the tomb of William of St Calais (dating to between 1080 and 1090) (see Pl. 9a, b), and the embroidery fragments from Worcester (dating to between the 11th and 12th centuries) (see Pl. 1b). Paul Garside has noted that wire thread makers knew how to make silver-gilt threads from the 10th century, while Plenderleith has observed that from the 11th century there was considerable debasement of the metal used to make wire threads, with the addition of copper or silver.[104] With the scarcity of gold, particularly during the 8th/9th to 11th centuries due to Viking raids and the paying of Danegeld,[105] we can hypothesise that these other forms of metal thread would have become more popular. Whether silver-gilt thread was produced in Anglo-Saxon England or imported is not known but it also became popular on the Continent, so it could have been sourced both locally and from overseas traders.

The silver thread used to create the embroidery discovered at Ingleby in Derbyshire is something of an anomaly. This thread was created by heating a silver bar and pulling it through holes of narrowing diameter until it formed a wire thread measuring 1 mm

---

102 E. Coatsworth and M. Pinder, *The Art of the Anglo-Saxon Goldsmith* (Woodbridge: Boydell Press; repr. 2010), 207–226; Hinton, *Gold and Gilt*, 88, 98–99, 165, 166.
103 Pinder also suggests the winding of gold around a core would be something fit for an apprentice: Pers. comm. Michael Pinder (2013).
104 P. Garside, 'Gold and Silver Metal Thread', in *Encyclopedia of Medieval Dress and Textiles of the British Isles c. 450-1450*, eds G. Owen-Crocker, E. Coatsworth and M. Hayward (Leiden: Brill, 2012), 237–239 (238); Plenderleith, 'The Stole and Maniples (a) the technique', 378.
105 Hinton, *Gold and Gilt*, 114.
106 Pers. comm. Michael Pinder (2014).

in diameter.[106] This is the only example of this kind of metal thread embroidery so far found in Anglo-Saxon England, whereas there are a number of examples of pulled metal thread woven into diverse objects from Viking sites in York and the Isle of Man, and other examples woven into decorative bands applied to textiles from the trading site of Birka in Sweden.[107] Agnes Geijer wrote that the method used to produce this form of thread was known to goldsmiths possibly working at Birka and may have been developed from knowledge gained from the Sami people who lived in northern Sweden.[108] This group used the same manufacturing method to make pulled wire thread out of pewter.[109] Given the situation in which this embroidery fragment was found, in the burial of a warrior from the Viking Great Army (see p. 84), it is likely that the thread was brought to Derbyshire from Sweden or another Scandinavian country and was not actually produced in Anglo-Saxon England. So while this piece of silver thread embroidery is an anomaly, it also contributes to the bigger picture of international exchange and trade networks established in the preceding discussion. At the beginning of the period embroidery was made from locally produced materials manufactured as required. Once larger estates and workshops evolved, they operated as part of wider trading networks, with materials bought and stored in bulk ready for use on particular commissions. Independent workers probably continued to buy smaller amounts that could be easily stored within a more confined space or relied on patrons to provide materials for a particular project.

Techniques (and working methods)

The technical construction of embroideries themselves is an important yet understudied source of information with regard to embroidery's production in the early medieval period. This involves not just choice of stitches, but how stitches have been formed, the order in which different sections of embroidery were worked, how the sections relate to each other, and the techniques of individual workers. Such details can divulge much about production of individual embroideries and the data can also be used to hypothesise wider theories concerned with standardisation of workmanship, level of stitcher training, whether a piece was domestically or

---

107 See R.A. Hall, 'A Silver Appliqué from St Mary Bishophill Senior, York', *The Yorkshire Archaeological Journal*, 70 (1998), 61–66; J.A. Graham-Campbell, 'Tenth-Century Graves: The Viking-Age Artefacts and their Significance', in *Excavation on St Patrick's Isle Peel, Isle of Man, 1982-88 Prehistoric, Viking, Medieval and Later*, eds A.M. Cubbon, P.J. Davey and M. Gelling (Liverpool: Liverpool University Press, 2002) 83–98 (88–89); A. Geijer, *Birka III: Die Textilfunde aus den Grabern* (Uppsala: Kungl Vitterhets Historie de Antikvitets Akademien, 1938).
108 A. Geijer, 'The Textile finds from Birka', *Acta Archaeologica*, 50 (1979), 209–222 (216).
109 Geijer, 'The Textile finds from Birka', 215, 216.

professionally produced, and the working practices employed. The analysis of embroidery technique that follows is arranged under three headings, worker skill, working methods, and workspace set up.

Worker skill

The skill of the worker can be ascertained by how stitches are produced – their neatness, the consistency of stitch length and the uniformity of the angle at which stitches are applied; and the level of coverage achieved for a particular element or motif. By analysing the technique of stitch construction it is possible to identify different hands at work. The Bayeux Tapestry, for which I undertook a detailed analysis of the embroidery work around the eight seams using photographs of the reverse, offers a good example. At seam two, which is covered by the death scene of Edward the Confessor, a very slight difference in stitch tension is identifiable between the work on the top floor of the structure (scene 27) and that of the ground floor (scene 28), the laid-work threads that overlay the base layer are more evenly spaced, and the embroidered outlines are neater, and more consistent in the top part of the scene, demonstrating that a different worker probably completed each section.[110]

The Ingleby fragment provides a rather different case. Here, the use of a solid silver thread would have given an inexperienced worker a number of problems. In order for this type of thread to be usable it first needed to be annealed, that is, warmed up, to make it pliable and soft enough to sew with.[111] The flexibility does not last long so the thread has to be worked quickly before it cools, loses its flexibility and snaps.[112] The wire thread may have been worked in short lengths to avoid breakages. The person who created the Ingleby embroidery must have understood these processes and been able to work adeptly within these limitations to create the finished item.

A precision and understanding of materials and technique can also be seen in the surviving exemplars of looped stitch. As alluded to previously, this stitch was worked throughout the early medieval period, with the earliest surviving piece, Orkney, providing one of the most complex variations (see Glossary). It makes use of three different threads worked at the same time to create the finished stitch. The variation shown in Glossary Fig. 68a is simpler in form and is found on objects dating from throughout the period, with the earliest being the 7th-century Sutton Hoo A.

---

110 A. Lester-Makin, 'The Front Tells the Story, the Back Tells the History: a technical discussion of the embroidering of the Bayeux Tapestry', in *Making Sense of the Bayeux Tapestry: Readings and Reworkings*, eds A.C. Henderson with G.R. Owen-Crocker (Manchester: Manchester University Press, 2016) 23–40 (30).
111 I am grateful to Michael Pinder for discussing this subject with me and pointing out these material qualities of high purity silver wire.
112 This is what happened in the experiments I undertook.

The other two examples come from York and Dublin, and date to the 10th century. The Orkney, Sutton Hoo and Dublin stitches all join seams together while the York version was used to bind the edge of a probable cuff. All four pieces were worked free hand as opposed to on a frame. This meant the embroiderer needed to multi-task: to keep the two pieces in play stable whilst, at the same time, weaving the thread(s) to create the plaited looped stitch; manipulating both pieces and threads with dexterity in order to work round the seam or edge evenly.

Each variant of this stitch work is very fine and the stitches have to be accurately placed in order for the next stitch to be sited correctly in the sequence.[113] Moreover, the use of the same coloured thread and fabric make it particularly demanding to work the stitch form correctly because the thread blends with the ground.[114] Although these seams and bound edges are modest rather than eye-catching, the worker needs to be just as technically proficient as those embroidering the gold and silk pieces in the later early medieval period.

It bears repeating here that nearly all the embroideries of this period are finely produced. This applies particularly to the length of stitches and the accuracy of discrete elements. The average stitch length throughout the period was 1 mm. To create embroidery that is not disjointed using such a small stitch, and achieve such flowing designs, is a feat that many professional embroiderers today would find difficult. That this was achieved throughout the early medieval period despite the lack of facilities enjoyed by professional embroiderers today, such as large windows, good lighting and ergonomically designed chairs, underlines the finesse on display. The earliest piece to exhibit such skill is the 7th-century Kempston fragment (see Pl. 2b), which is worked in wool threads on a worsted wool ground fabric. The same high standards can be seen on later pieces in metal threads and silk. This is particularly so with the Ushaw fragment of Durham D. The surviving letters worked on this fragment are sewn in stem stitches measuring 1 mm each (see Pl. 17c and 20c). For motifs like letters a small stitch can sometimes be helpful as it enables a high level of definition. However, a letter containing curves can look scrappy if not worked correctly, so the embroiderer needs technical skill and an understanding of the materials in order to creative a cohesive effect.

---

113 The author's recreations measure 3 mm long per stitch by 4 mm wide. It was found that the stitch structure demands to be worked diminutively – when one works it in larger form one find that it has shrunk by the time the line of stitching is complete.

114 As participants who undertook to recreate the author's exemplar at the European Textile Forum Conference in November 2018 discovered, it is easy to misplace stitches and lose your place within the loop pattern if you lose concentration. They also found that the stitch is virtually impossible to construct over a seam unless the turned under raw edges of the fabric are hemmed in place, as per those of the Sutton Hoo piece.

The precision of the Durham embroideries shows that these embroidery-workers were trained to the highest levels. In particular, the consistency of stitch length and the evenness with which the complex couching of the gold was completed demonstrate these skills. These pieces were created by professional workers at the height of their powers for a patron who was able to pay for time and expensive materials (see Durham C, D and E).

On the other hand, the slightly earlier Maaseik embroideries are not as consistent as the Durham pieces. Stitch lengths measure between 3 and 5 mm. Overall, the workmanship is not of as high a standard as that seen on other gold and silk examples (such as Durham C, D and E). The execution appears less exceptional, although it is still of a good, consistent standard. The idea seems to have been to focus on filling areas as opposed to achieving the high degree of detailed decoration seen on Durham C and D. On the other hand, when compared to the exacting rigour of the Durham embroideries, the Maaseik roundel and arcade strips have a fluidity and 'life'. Moreover, while the fragment of Durham embroidery that I have been able to analyse in detail is outstanding, the difference in technical workmanship could simply be attributable to different stipulations and priorities on the part of the patron. Someone of royal lineage would have had more money to spend on the production of a commissioned piece, meaning more time could be allocated to the project. Given ample time, the workers need not have hurried the embroidery, and would have been able to produce it to higher standards.

Variation in how the stitches were used and the use of colours also contributes to the character of the Maaseik pieces. For the arcades (Pl. 23), the outlines of the gem-like motifs were picked out on the ground fabric using a red thread (now faded to pink). They were re-outlined in stem stitch with the coloured silks and then filled in, again with stem stitch (Pl. 24a). This method of working helped maintain the original shape of the motif throughout. For small motifs, such as triangles, stitches follow the edge of the shape at decreasing sizes to create the infill (see Pl. 24b). In other cases the stitching has been randomly placed to fill the motif (Pl. 24c), while curved areas have been worked in corresponding lines (Pl. 24d). The use of colour on the arcade strips incorporates bands of different colours worked within a single motif (see Pl. 23d). The use of colour is more complex on the roundel strips, where some motifs were worked in a single colour, others in blocks of different colours (Pl. 25a, b) or incorporating a single stripe of a different colour. These varied treatments of different areas could be an indication of different workers being responsible for embroidering the arcades and the roundels. Moreover, the couching is not as neat on the arcade strips as on the roundel bands, which may point to a lack of experience on the part of the worker(s), although it could also be a function of the shape of the motifs being worked. Interestingly, there is some variation between the two roundel strips as well, which may suggest the work of different embroiderers. For instance, the way stem stitch has been applied (all stitches facing in one direction, or alternatively, placed to form a chevron pattern) seems to follow different schemes on the two strips (see Pl. 25a, b and explanatory captions).

As we might expect, the couched gold thread on the Maaseik embroideries is not as technically exacting as that seen on Durham C, D and E either. With regard to the roundels, the couching has simply been worked to make sure the gold thread fits the shape of the design, with the couching threads crossing over the gold in lines (Pl. 25c), or sometimes in a more haphazard fashion (Pl. 25d). The gold on the arcade strips has been couched in either single or double lines (Pl. 26a). By contrast, the halos on the stole at Durham show a high degree of technical dexterity in the use of different patterns to hold the gold in place.

The way the gold is arranged also shows variation. Where the gold thread lies in straight lines the threads tend to be placed close together. Where the gold is laid around sharp corners it is, on the whole, bent with precision (Pl. 26b). However, when the gold is couched over a more fluid shape, the rows are not worked as closely together and the bends in the gold tend to occur further apart (see Pl. 25d) which suggests that the worker(s) were not as adept at filling these sorts of spaces with gold-work. Where similar shapes occur on Durham D, on the other hand, they are more densely filled with gold thread. However, arguably the shapes on Durham D and E are simpler to execute in gold thread, and the workers of the Maaseik embroideries may have been used to stitching simpler motifs in horizontal and vertical lines only, as opposed to adapting their work to suit more complex shapes. Again, inconsistencies in the Maaseik workmanship is indicative of a number of different embroiderers working as a team to complete the bands.

Working methods

By studying the order in which an embroidered item was created, it is possible to gain insights into working methods. For instance, my previously published analysis of photographs of the reverse of the Bayeux Tapestry enabled me to assemble a detailed hypothesis regarding how the embroidery-workers tackled these areas, which in turn led me to deductions regarding workspace and project management.

It became clear that all embroiderers were trained to a similar professional standard (even allowing for slight differences in technique, discussed above). They were able to work discrete areas within the design as needed, as opposed to completing elements or motifs as complete entities. This gave greater flexibility, enabling them to move between sections as required, or perhaps specialise in particular aspects of the design, for example buildings or horses. It also meant that the wool embroidery threads were used economically.[115]

---

115 For the full analysis and discussion see Lester-Makin, 'The Front Tells the Story'.

We begin to appreciate, here, that the number of embroiderers and possibly the workspace layout would differ for an embroidery worked consistently from one side to the other and/or from top to bottom, as was probably the case with the fragmentary gold-work from Milan, compared to a piece where the work was split into smaller sections like the Bayeux Tapestry, or pieces like Durham C, D and E, which were worked as separate pieces that were sewn together on completion. In the first scenario, a worker would set up her workstation and continue working there until the embroidery was complete. If the embroidery being worked was large for the slate frame to which it was attached, the ground fabric could be wound round two of the four bars that made up the frame to which it was fastened (either top and bottom or at the sides). In such cases only the area being stitched would be on display. Once this section was completed, the tension of the ground fabric was loosened, the two sides not wound round the frame were detached, and the completed section was then wound round one bar, while the next area to be embroidered was unwound from the opposite bar. The ground fabric was then re-attached to the sides of the frame and the whole pulled taut. The worker would then begin embroidering a new part of the design. The process was repeated until the whole of the embroidery was complete. The fabric and completed sections of embroidery were protected by inserting a soft cloth, and winding it round the bars and in between the layers of completed rolled-up embroidery, to prevent stitches from rubbing against each other, or the ground fabric (see Fig. 46).

In the second scenario, which is a plausible proposition for the way the Bayeux Tapestry's complex architectural motifs were produced, workers moved from frame to frame, completing sections as required.[116] The third plausible working arrangement entailed one or more embroiderers working on different self-contained parts of a commission at discrete locations within a particular workspace, or local geographic area. Once all the sections were finished, they would be brought together to be assembled as a finished object. Both the second and third methods of working would be useful for more complex large, flat pieces of decorative or narrative embroidery.

The Maaseik pieces also allow us to distinguish aspects of a clear working method. The design was laid out on the ground fabric using a single strand of red silk thread.[117] Once this was complete the silk embroidery was worked. The evidence for this is provided by a number of silk embroidery stitches passing over or piercing the outline thread (Pl. 26c). Once the silk-work was complete, the gold thread was couched in place. This is evidenced by couching threads catching the silk stem stitch infill and outlines (Pl. 26d). The gold appears to have been laid out in the same manner as the silk: from the outer edges of the motifs inwards. This would help the embroiderers to keep to the shape of the motif being filled. Indeed, it is still the method used by

---

116 Lester-Makin, 'The Front Tells the Story', 37.
117 Budny and Tweddle, 'The Maaseik Embroideries', 76.

professional embroiderers today. What we cannot tell with regard to the Maaseik embroideries from the available evidence, however, is whether the whole strip was embroidered from start to finish or in smaller sections like the Bayeux Tapestry. However, microscopic photography indicates that sections were worked as blocks, with coloured threads used as needed. As the mosaic form of the background involved the use of many different coloured threads, it is not unlikely that one gem-like section was stitched and the thread was then left hanging loose until it was needed again. This was the same working method employed on the Bayeux Tapestry, where threads were left loose over small distances ready to complete an element in the design.[118] Something akin to this is visible around the edge of the Maaseik embroidery.

The final example discussed here highlights a different aspect of working practice: the use of experimentation. The Worcester embroideries date to the end of the early medieval period and possibly later. Despite difficulties obtaining adequate photographic reference shots,[119] careful study of the front of the actual embroidery and analysis of photographs taken by Miki Komatsu during her MA research allowed me to confirm, contrary to previous researchers' assessments,[120] that all the metal thread work is underside couching (where the couching thread is pulled through to the reverse of the embroidery so that it cannot be seen from the front, leaving a small indentation in the line of the gold thread being stitched in place) (see Glossary). This technique was used to render the hair and clothing of figures and upper and lower bars of the frame. The line of the metal thread lies in a slight diagonal, from the top right to bottom left of the clothing on the figures facing right, and from the top left to bottom right of those facing left. The underside couching creates small indentations in the metal thread on the front of the design, in a consistent pattern of slightly curved horizontal lines (Pl. 26e, f).

---

118 Lester-Makin, 'The Front Tells the Story', 24–25.
119 Analysis of technical aspects of the pieces was hampered when viewing the embroideries in several ways. First, by inability to view the reverse of the textiles: they sit on conservation fabric within a pressure-sealed frame; second by the faded brown colouration of the ground fabric and embroidery threads; and thirdly by the low lighting in the cathedral library, which is also an impediment to research. In order to gain a better understanding of how the embroidery was executed, I examined all the pieces visually and took photographs and microscopic images. However, reflection from the frame's conservation glass meant they were not successful. As a result, I had to rely on the images taken by Miki Komatsu during her MA research and are now held at the Textile Conservation Centre, University of Glasgow.
120 Christie, *English Medieval Embroidery*, 52, 53; M. Komatsu, 'Investigation of the Fragments of Liturgical Textiles from Worcester Cathedral' (unpublished MA dissertation, University of Southampton, Textile Conservation Centre, 2007), 51. These researchers stated that the side bars of the frames on the stole and maniple, the whole frame on wedge 14 and some of the frame on wedge 16 were embroidered in feather stitch or false satin stitch; my research showed it was actually underside couching used to create a feather-like design, not a different stitch. Komatsu also argued that the lettering, parts of the hair and the dots were worked in false satin stitch but images of the reverse actually show small, and in some cases very tightly formed, loops of metal thread which are the same as those found to the back of areas covered with underside couching.

The side bars of the frames, the sword handle, lettering and the halos were all worked by laying the metal thread at different angles, as opposed to in rows next to each other, to create the design required. The metal thread was then underside couched in place at each turning point in the design, creating small lines of metal instead of long continuous rows (Pl. 27a, b). Using this technique allowed for more complicated shapes to be formed. It has also been used to create the frame that surrounds the animal motif on Durham G. The dots, which measure 2 mm wide and between 1 and 3 mm high, are worked in the same manner as the letters. Each circle contains between 4 and 9 lines of metal thread underside couched in place at each end of the rows. The metal thread is then bent round to form the next line so it lies directly next to the previous one. The length of each row is altered alternately at either end so the shape of the dot is formed (see dots to the right of the figure in Pl. 30a).

Underside couching is rare on most surviving embroidery from early medieval England or Ireland. There is a single fragment from Durham (G) (see Pl. 9a) which dates to the end of the Anglo-Saxon period. It is likely that originally there were more embroideries that utilised this stitch, but no more are known to survive at present.

The details rendered in silk thread are also in what we might class as unusual stitch forms. The only surviving pieces with comparable designs are Durham C and D. The details on the Durham stole and maniple were worked in split stitch.[121] This was also the technique used in extant examples of the later *Opus Anglicanum* embroidery (Pl. 28a, b, c).[122] I would argue that the areas of split stitch found on the Durham embroideries are a development towards *Opus Anglicanum* because the stitching was worked both vertically, for example across the faces, and horizontally, on the foreheads and necks. Examples of *Opus Anglicanum* show a more realistic curvilinear use of the stitch, for example the shape and muscles of the face were rendered through the stitch work; circles were used on the cheeks and ovals across the forehead and neck. The Worcester embroideries do not fit into this progression because the stitch work is vertical throughout the features, the only exceptions being the eyes, where horizontal lines of stitching have been used instead of split stitch, probably to add emphasis to these areas. The shape of the eye was filled with silk thread laid horizontally across the space. At each end it was turned and brought back across the eye, and forming the next stitch. The thread was held in place at each turning point with a small holding stitch.

The figures on the Worcester fragments are not of the same style as those that decorate the Durham embroideries (see Pl. 28a, b). The Durham figures are very much of the 10th century, while those on the Worcester embroideries are much stiffer, with simple clothing. As such they do not fall into the same category as those found on *Opus Anglicanum*. Interestingly, they do resemble some of those from the Bayeux Tapestry:

---

[121] Plenderleith, 'The Stole and Maniples (a) the technique', 380–381.
[122] Christie, *English Medieval Embroidery*, 2–7.
[123] Komatsu, 'Investigation of the Fragments of Liturgical Textiles', 33.

for example, the bishop on one of two wedge-shaped panels looks remarkably like Archbishop Stigand in scene 32, where Harold is crowned king (Pl. 29a, b). The crowns of the kings and the sceptre they hold are similar to Edward's crown and sceptre in the opening scene of the tapestry, and the stance shown by the apostles can be seen in various places throughout the Tapestry (Pl. 30a, b). It has previously been noted that the buildings in the arches of the wedge-shaped pieces are like those of the hanging,[123] particularly, I would suggest, the building on fire in scene 51 (Pl. 31a, b, c, d). Moreover, the way the clothes of the Worcester figures have been designed and stitched is similar to the clothes of people populating the Bayeux Tapestry. This can be seen in the sloping shoulders and stance of the figures; the way the arms have been drawn, bent at the elbows, sometimes at awkward angles; the long lines of the tunics; and the embroidery, which gives the garments comparable textures through the laid-work on the Bayeux Tapestry and the indentations of the gold thread on the Worcester fragments (see Pl. 30a, b). It can therefore be argued that the Canterbury Style was an influence on the designer of the Worcester embroideries. The designer and workers may also have been influenced and/or trained in this style as well. However, the layout of the figures on the Worcester embroideries is the same as those on other stoles and maniples with each figure being framed in some way. Indeed, framing scenes or individual figures can be seen throughout the medieval period with items worked in *Opus Anglicanum* using elaborate frames of arches, simple versions of which can be seen on the wedge-shaped Worcester pieces. The figures of the stole and maniple hark back to the Durham embroideries. This suggests they are a very early phase in the development of *Opus Anglicanum*. The fact that they are cruder than the fine work produced previously points to the early stages of development of a form that is feeling its way towards a new style.

Although the surviving pieces are too fragmentary and small in number to allow for definite conclusions to be drawn, a hypothesis about working methods can be posited. The stole has survived in better condition than the maniple, and the two wedge-shaped pieces have survived least well.

Nine apostles survive on the stole: six facing right and three facing left. The group of six show a number of similarities which suggest features, and indeed figures, being worked by the same embroiderer, for example, the head and halo of two apostles exhibit the same form, particularly along the top of the hair-line, around the chin and neck and where the halo passes behind the neck. Furthermore the feet and bent arm of these two figures are very similar in design and technical style. Stylistic similarities are also visible with regard to two other apostles and again between a third pair. Of those apostles facing left, two of the three appear to share common design and stylistic features in the shapes of halos, hairlines, chins and feet.

It would thus seem plausible that the two sides of the stole were laid out separately and worked simultaneously by different hands, with at least two working on each side. If more workers were involved, this would lead to the assumption that each figure was produced as an independent section in a similar manner to the Durham stole and maniple, with particular figures being embroidered by different teams of workers, but the lack of extant segments means that we do not have sufficient evidence to flesh out such a hypothesis. While too fragmentary to permit a detailed comparative exercise, tantalisingly, the remnants of one figure on the maniple appear to be more accurately rendered than the apostles on the stole, suggesting another discrete group of workers or a worker whose embroidery does not survive on the stole was employed to embroider these figures.

Workspace organisation in the later early medieval period

Following on from the detailed discussions above, we can hypothesise particular *modus operandi* for the different types of workspace used for each embroidery.

Turning first to the Bayeux Tapestry, as discussed in 'The Front Tells the Story', the consistency and economy of working practices displayed across the hanging, the standard of technical skill and the evidence for teams of workers completing designated parts of the design all point towards the presence of a quality control manager overseeing the work. In other words, we can infer that labour was being directed and managed, and a workroom manager, skilled at organising and monitoring the work, was most likely in charge of the whole process. This indicates that the whole hanging was worked in a single large workshop or several workshops located close to each other (travelling large distances between workrooms would make control of the production process harder, and is unlikely to have been the case).[124] The conclusions are thus quite radical, pointing to methods of working with similarities to those of the modern embroidery studio.[125] Workers may have been drawn from across the country, and housed and working together like the previously discussed goldsmiths employed by King Edgar.[126]

The Durham headdress ribbons or girdle, stole and maniple (Durham C, D and E respectively) offer tantalising insights into a second viable workshop arrangement. As already discussed, the stole and maniple were embroidered in individual sections, each figure being worked separately before the finished pieces were sewn together. Evidence of a similar working method is also apparent for the third piece. As discussed previously (pp. 14–15), Elizabeth Coatsworth has persuasively argued that this piece was probably not initially intended as the single item we see today but was more

---

124 Lester-Makin, 'The Front Tells the Story', 37.
125 Lester-Makin, 'The Front Tells the Story', 37.
126 Lester-Makin, 'The Front Tells the Story', 36, 37.
127 Plenderleith, 'The Stole and Maniples (a) the technique', 390–391.
128 Although there is no evidence that the design was further divided into smaller subsections undertaken by different workers (like stole and maniple) but a microscopic study of the piece may prove otherwise.

likely used as two separate ribbons for a headdress, which could possibly have been joined at what is now the top, to form a girdle. Although it is embroidered on both the front and reverse, Plenderleith's analysis found that the stitching was not worked on the two sides at once (as in the 'both sides alike' technique, used for military banners).[127] Instead, the piece was embroidered as four individual sections, two for the front and two for the back, which were sewn together afterwards to create a reversible embroidery.[128] The different sections would have been attached individually to separate, suitably sized frames. Each complete section measures approximately 636 mm long by 24 mm wide, so those organising the project could choose to use small frames which necessitated some winding of excess fabric round the bars of the frame, or larger frames which would not require winding excess fabric but would need more space in the workroom (slate frames can measure 1 m or more in length today and large frames may have been used in the early medieval period as well). Assuming more than one section was being worked simultaneously, a large room or even a set of rooms would have been needed to accommodate the frames (particularly if the larger ones were used), multiple workers, and space for people to move around and store the materials and equipment for a considerable period of time.

This leads to the conclusion that a workshop setting not that different to that hypothesised for the Bayeux Tapestry may have been utilised, or, since at least the stole and maniple were commissioned by Queen Ælfflaed, and since the third piece was obviously made by the same workers (see p. 130), it is possible that the workers were gathered together, housed (possibly on an estate), and provided with materials at the queen's instigation. If the ribbons/girdle were framed up on smaller slate frames, the separate sections could conceivably have been embroidered in smaller discrete locations by a dispersed work-force working under the guidance of a manager like Ælfbold of Thetford (see p. 116). However the three embroideries are so similar, with the four sections of the third piece technically indistinguishable, that it is highly likely the embroiderers were working in close proximity, trained to a set standard and organised by a manager – either in a single large workshop, or a group of slightly smaller workshops located close together.

The Milan embroidery points to a third alternative working environment. This small, now fragmentary piece of gold-work was originally stitched on to a silk ground fabric and was likely stitched by a single worker using a smaller sized slate frame. It is the sort of work that could have been undertaken by an independent worker such as Aelfgeth and Leofgyth in Domesday Book or Liveva and Ingrith in the *Liber Eliensis*. The materials – silk fabric and thread and gold metal thread – were precious and expensive so although the embroiderers could have purchased and stored them for future use, it is more likely that they were supplied with the materials by the patron when the work was commissioned. Independent workers would have worked in their own homes or small designated workspaces on their *lænland*. Although there would be no workroom manager supervising quality here, the patron may have made visits to view the work in progress, and this, together with self-monitoring, would have helped to ensure work was completed to a consistently high standard in order to keep the commissions coming.

The Maaseik and Worcester embroideries appear to present variations of the larger workshop arrangement. The inconsistencies in workmanship discussed with regard to the Maaseik embroideries are another form of evidence for collective working practices. They are suggestive of a number of different embroiderers working as a team to complete the bands, each probably with agreed quantities of work and particular areas of focus. This, combined with evidence of different parts of the work being produced from different batches of material, suggests that groups of embroideries were being created simultaneously and embroiderers were expected to work on whichever parts needed attention.

Finally, the Worcester setup indicates a workshop environment where commissions for sets of embroideries – such as a stole and maniple – were taken on and worked as an integrated project by those employed there. In this case, all the workers would likely have been trained to a similar standard in most forms of popular embroidery techniques. Moreover, it can be hypothesised that the Worcester pieces were produced in a workshop setting by teams of trained workers tasked to produce each set of embroideries. The use of silver-gilt thread and stitches such as underside couching dates to the very end of the Anglo-Saxon period and beyond. The execution is not as accomplished as the work seen on extant examples of *Opus Anglicanum* created during the later period when the use of underside couching was at its peak. Rather, in its use of technique and materials the Worcester embroideries can be seen as part of the development towards the *Opus Anglicanum* style. At the same time it shows design affinities with older works: the rendering of outlines in stem stitch found on Durham D and E, and the style of buildings seen on the Bayeux Tapestry. I would therefore argue that the Worcester pieces involve experimental work practices, for both the underside couching and the silk-work lead to the hypothesis that this embroidery is an important piece in the evolution from early medieval to *Opus Anglicanum* embroidery.

This study of technique has shown that embroiderers' hands, skill, knowledge and thought processes can be ascertained from their work: from merely filling spaces to the decorative, and from understanding the process of the materials being worked to pushing forward what can be achieved with them through experimental work. The working methods can give researchers clues as to the flexibility of working environments and their layout, from a basic one worker per frame to a team working on numerous frames, or specialising on certain motifs within a project. Analysis of technical aspects of embroidery construction thus helps us to understand wider issues concerned with the organisation of Anglo-Saxon production.

Conclusion

The material deployed here is used as evidence to propose a pattern of development in embroidery production over the course of the early medieval period. At the beginning of the era, embroidery was being worked by female members of settlements at a local level. Training was probably informal with mothers and elders teaching younger generations. Materials utilised were produced locally, and the surviving archaeological evidence suggests this was mainly wool.

During the middle period, as estates developed, while the general populace continued to learn to embroider and undertake embroidery work as before, a more structured approach was beginning to develop. Elites and the Church were demanding more specialised goods worked in a wider variety of materials, including linen and prestigious silks. Estates such as the one run by Eanswitha, are likely to have had embroiderers working for them and may have offered a form of apprenticeship or school in which to train others. Some materials such as wool and flax would have been gathered from the estate and stored for when needed. Other products were probably obtained from traders either through exchange or trading networks.

As the population became more urbanised, patterns of working established on estates solidified into workshop settings, perhaps with workers training on apprenticeship style schemes before graduating to work in a workroom. These were highly proficient workers who were equally at ease completing individual pieces and sections of larger embroideries, or working more flexibly on particular motifs within a design. The level of technical skill required for this, alongside manipulating materials, perhaps contributing to designs, and working in such a way as to match individual technique to the type of work required in different areas of the same piece for maximum impact, should not be underestimated.

Independent embroider-workers perhaps took the opportunity to set up small cottage businesses in urban areas, possibly taking up part of a room in a dwelling. In towns such as York, Dublin and Winchester, they may have been situated in shops that fronted onto the street with living quarters behind. If the worker also engaged in teaching like Ælfgyth did for Sheriff Godric's daughter, she may had had a separate room or building in which she both worked and taught. These embroiderers would have produced work to commission. Like Alderet's wife, it is possible that they had a particular patron or a number of patrons. Alternatively, they may have undertaken piecework for an intermediary such as Ælfbold, the burgess.

While workshops would have been able to buy and store larger quantities of the materials required, small independent concerns would have probably relied upon the patron to provide the materials. Workers may also have purchased smaller amounts of thread and fabric from local traders or other individuals, such as enterprising goldsmiths. In particular, women of higher rank who lived in small lay communities would have had this sort of purchasing power. Women of elite and royal circles would most likely have received materials through gift exchange or been able to purchase those they needed through diplomatic trading channels, as well as utilising the resources of their own estates. Such women would have been taught to embroider from an early age, probably by their family and women of the court. Others attended convent schools and were taught their craft as part of a broader educational programme. In turn, they may have passed on the skill to young girls within their circle. As well as embellishing the clothes of family members, as Edith did (so we are told), women were able to use their skills to allow them to act as knowledgeable patrons of the Church, as Queen Ælfflæd did when she commissioned the stole and maniple for Bishop Frithestan (Durham D and E) or as Queen Matilda did when she commissioned the chasuble from Alderet's wife.

This, then, is the story of the development of embroidery within Anglo-Saxon society, and of the development of an important, accepted form of female employment. By the later stages of the early medieval period, once trained, an embroiderer could work on an estate, in a workshop or independently. What began as one of women's roles within the individual settlement became part of an integrated production and training system that produced work prized throughout early medieval Europe.

In the final chapter I will discuss how embroidery can be understood as part of broader patterns of change: as developing to fulfil the needs of an increasingly complex society. The desire for embellished material culture allowed women to carve a place for themselves within society. As the need for more elaborate embroidery developed, enterprising workers took advantage.

Chapter 6

Conclusion: embroidery in context

Introduction

This project set out to establish the position of embroidery in the early medieval British Isles and Ireland. It recognises the pitfalls and difficulties involved in studying this fragile and ephemeral form of material culture. Yet it has shown that an interdisciplinary approach, using technical data and contemporary sources, together with other art forms as a comparative framework, provides an effective methodology. Most importantly, the use of object biography theory shows how painstakingly assembling the biographical detail of what we see, today, as a small innocuous embroidery can help to establish a wider, contextual story of those who were involved in its creation, use and deposition across multiple lifecycles.

Similarly, deploying analysis of technical data in combination with object biography across a range of items helps to draw out and demonstrate that embroidery was a significant and integrated part of Anglo-Saxon material culture. The interdisciplinary analysis of extant embroidery, documentary sources and other art forms effectively tells the story of the development of embroidery production in Anglo-Saxon society and the pivotal role of women as embroiderers, without romanticising these concepts, nor projecting our own values onto a society with such different ways of understanding the world.

The significance of embroidery in early medieval society

Social and material contexts

Embroidery played a pivotal role as a material expression of early medieval society. It was not just a form of decoration or artistic expression; discussion has demonstrated that it was also an important political and religious tool. Throughout the period, embroidery was used as a marker. During the earlier centuries, when the Germanic tribes first settled in Britain, individuals used embroidery to identify initially with their ancestral homelands and later with their new kingdom and tribal allegiances. This can be seen particularly well in the form of the 7th-century embroidered cuff from Mound 14 at Sutton Hoo, which mimicked metal cuffs found in Norway and East Anglia. In each of these locations the metal cuffs took on new, subtly different roles as expressions of both gender and affiliation, and I have argued that the embroidered versions are a progression of this. Such markers were not only reminders for those wearing them but also messages to those who saw them. They told the onlooker not only about the wearer's ties but imparted symbolic power and political strength as a result of these tribal links.

As the period progressed and smaller principalities were swallowed up by the more powerful Anglo-Saxon kingdoms, embroidery continued to be used to demonstrate powerful associations. It was now also possible, however, to identify the wearer with more outward-looking inter-territorial values and ideas; for instance, the late 9th- to 10th-century embroidered Llangorse textile appears to mimic the woven silks of imperial Byzantium. The wearer of such a garment would thus be expressing the trappings of 'imperial' power to his/her subjects and aligning him/herself with the status and authority of the Christian Byzantine rulers.

As kingdoms merged in the tenth century to become Anglo-Saxon England, middle and elite ranking individuals used clothing to express status, using the highest quality textiles they could afford on objects such as the decorated vamp of the child's shoe from Coventry. Embellishments such as embroidery would also have been valued. During the later part of the period women embroiderers of rank donated embroidered items to the Church and its saints to show favour and patronage to the ecclesiastical elite, as well as to ask for God's help. The early 10th-century Durham stole and maniple are examples of embroideries functioning in both these ways. With these two pieces their commissioner, Queen Ælfflæd, was showing favour to Bishop Frithestan, as well as serving and honouring God by providing the Church with much needed vestments. The embroideries took on a new role when King Æthelstan, Ælfflæd's stepson, took and deposited them in the coffin of St Cuthbert. At this point the embroideries were being used to ask for specific intercessions of the saint, to help the king in his fight against the Picts. Such a sumptuous gift also asserted political ties between the king and the monks.

As one of the last truly Anglo-Saxon embroideries, the 11th-century Bayeux Tapestry provides a final example of an embroidery with political import. In this textile the once mighty Anglo-Saxons could read the story of their downfall and subsequent domination by the Normans. No matter who commissioned this work, it was a powerful statement of political and religious change, and of power and strength in a new guise; a final message of new relationships, ties and authority.

Embroidery and meaning: early medieval mindsets

As we have seen, during the early part of the period, embroidery appears to have been imbued with an additional level of metaphysical meaning. The clearest example of this is looped stitch. At a functional level this stitch was used decoratively to reinforce weak points on textiles, but in its form it may also have represented the World Serpent of Germanic mythology. As such, the stitch would have been intended to take on attributes commonly associated with the snake, for example, its protective nature, and the use of this stitch on areas susceptible to wear and damage was more than simply functional: it had symbolic power. The use of looped stitch coincides with the period of belief in pagan deities; from the 6th century, once the Anglo-Saxons had converted to Christianity, its use appears to have waned. With the arrival of the pagan Scandinavians from the 8th century, however, the stitch was re-introduced to England; they continued to use it to decorate, reinforce and protect their textiles. By this time, use on objects owned by the elite ranks of Scandinavian society had helped the stitch evolve and take on a new form worked in pulled metal wire. The Ingleby piece suggests that elite warriors expressed their status and beliefs in items constructed from this new, difficult-to-manipulate thread. At the same time those of middle rank, such as traders in Dublin and York, continued to utilise the stitch's protective nature in the traditional material form of wool.

Within Christian religious settings the Anglo-Saxons used embroidery to express multiple messages; but not, it seems, to enact the message or act itself in the way looped stitch did. Embroidery such as the 10th-century Cuthbert stole and maniple was now seen as a material conduit between the earthly sphere and the heavenly, through the intercession of the saints and the prayers of the clergy. Within the earthly realm, pictorial representations of saints, popes and holy men were reminders to clergy of their sacred role within the Church, and of how they should live their lives. These same images were used to impress and inspire the congregation with both the authority of the Church and its role as a moral exemplar, helping people to lead better, safer lives in a world of harsh and dangerous reality, unreliable environmental conditions, famine, death, violent inter-kingdom wars and Viking raids.

Embroidery could also embody religious intentions at an individual level. The York pouch is an example of this. This bag, made from a scrap of precious silk fabric, was embroidered with a Latin cross worked in silk thread. It is small and personal, to be worn around the neck of the owner. Inside it most likely held a relic of a saint. God's power could be transferred by the holiness of the saint, through their relic, to the individual who wore it, thereby protecting or benefiting the wearer at a private, personal level. In this case, embroidery was not the metaphysical form of protection afforded by looped stitch, but a material conduit through which the new form of religious help could be transferred from one realm to another. By virtue of its representation of the Christian cross, the embroidery on the pouch was a form of message, transmitted privately, at an individual level, in contrast to those on

Church vestments, which communicated such messages on a grand scale within a public arena.

In one of the final pieces to be created during the period, religious message-carrying can be seen operating in new, experimental form, not in terms of the design, which was quite traditional, but in the use of materials. The late 11th- to 12th-century Worcester embroideries continued in the same design tradition as the Durham vestments; however, the elaborate nature of the stitching and use of metal threads – worked in underside couching instead of surface couching – can be seen as an evolution towards the later *Opus Anglicanum* style. The overall intention was the same, but the means of expression had evolved. The use of contoured metal thread, bending towards and through the ground fabric, would have helped to project a newly effective image of the Church's power as it flickered and reflected in the light from the candles. The elaborate nature of this stitch work, created to honour the radiance of Christ as 'Light of the World', was also an artefact of the imagination of the people who created it, pushing the boundaries of what was achievable with the precious and vulnerable materials with which they worked. Understood from a religious standpoint, the pieces can be seen as a form of praise through work; the embroiderers were worshipping God by using the talents he had given them.

The production of embroidery and its makers

The evolution of production processes can be seen as one of the defining factors for the development of embroidery as a form of textile decoration. As the organisation of society changed, so did the production of embroidery. Between the 5th and 7th centuries embroidery was created within the settlement, as part of the everyday life of the women and girls who produced the textiles needed by inhabitants. Locally produced fibres were used, and if they were dyed, this was done so in the same vicinity. Work took place within the home or under external canopies. In Ireland the evidence suggests that particular locations were set aside for such work. Embroidery was used to embellish textiles that were produced by members of the settlement, at least for the local populace. However, even at this early date, the evidence from Ireland suggests embroidery was seen as something special, to be undertaken by particular members of the community, presumably those women with expertise in the techniques.

As settlements evolved so too did the production of embroidery, particularly with the development of central locations (larger estates) during the 7th century. Now embroidery not only decorated the clothes of villagers and the local elite, it could also be used to embellish textiles that were to be traded or exchanged for other goods. We can hypothesise that, in order to gain a better return on each trade, there was pressure to raise the quality of textiles and embroidery. Combined with the growing demand from the newly established churches, this must have led to a development of occupational specialisations, both in the production of embroidery and the training of workers. Such advances may be seen in the estates run by Eanswitha and the possible school of Æthelswith. Within these larger estates batches of fabrics could be made and stored with larger quantities of thread – either for use when needed or ready for trading – while smaller concerns would either be supplied with the necessary materials by patrons or could opt to trade or buy in the fibres in limited quantities as and when they were needed. Many estates, like Eanswith's, were run by the Church. Others were organised and controlled by ruling families who were in charge of running swathes of land. Thus both the Church and these families played an instrumental role in the gradual reorganisation of society and the nurturing and development of more specialised occupations among those they ruled. More diverse trading opportunities emerged as a result of the links between these groups, leading to the importation and production of new materials that could be incorporated into embroideries. This in turn allowed the Church and the elite to demand exotic fibres such as silk, in newly produced textiles, as opposed to relying on textiles received through gift and diplomatic exchange. Silk thread could now be combined with the precious metal threads produced by goldsmiths to create embroidered bands or cover complete garments with stitch work.

As the population became more urbanised, particularly from the 9th century, it became more common for women to set up independent embroidery establishments in centres such as York, Southampton and Winchester. Now these embroiderers could take on commissions from anyone who could afford them, whether from religious institutions, or from elite or middle ranking individuals. Such examples can be seen with Liveva and Ingrid at Ely, Alderet's wife in Winchester and Ælfgyth in Buckinghamshire. The first two women probably took on commissions from the monks at Ely, while Alderet's wife is known to have worked for those of elite ranks, including William the Conqueror's wife, Matilda, who commissioned textiles from her. She may also have taken on church commissions, since Winchester was an important ecclesiastical centre needing vestments. Ælfgyth seems to have taken on secular teaching as well as commissions, as it is for just this reason that she was mentioned in Domesday Book. All of these women were highly regarded, with their finished embroideries being prized and commented upon in contemporary sources. It is likely that their work was re-used and recycled through multiple lifecycles until it was no longer perceived to be of use by owners. Alternatively, it was purposefully taken out of circulation, as in the case of the late 8th- to early 9th-century embroidered Maaseik bands, when they were given to the shrine of Sts Harlindis and Relindis.

The technical analysis of a number of the later embroideries, including the 10th-century Milan fragment and the 11th-century Bayeux Tapestry, shows how embroidery workshops had also developed by the end of the period. In particular, the stitching of the Bayeux Tapestry demonstrates that, as well as being used to working whole embroideries, women were adept at focusing their efforts on certain motifs or set elements within a design as needed. Whilst it is probable that some women were continually employed in workshops, it may also be the case that, when the need dictated, greater numbers were called in to work on larger projects, as has been documented for goldsmith work. It is also possible that women were trained in the art of embroidery by a workshop and then went on to work semi-independently, on the understanding, perhaps, that they would be called upon for these larger projects when required.

So here is evidence that, while some worked as independent embroiderers, there were also the beginnings of a workshop system. Indeed, it may be that Ælfbold, the burgess of Thetford mentioned in the *Liber Eliensis*, ran such a fledging workshop. From these establishments the creation of apprentice-style schemes must surely have emerged in order to maintain the standards and technical skills required to create valuable, prized embroideries, and which would eventually lead to the development of *Opus Anglicanum*. Moreover, while women from secular society were developing entrepreneurial opportunities, so too were convents such as Wilton, which was located in the Anglo-Saxon kingdom of Wessex. These were educational establishments for women of high rank teaching a range of subjects, including embroidery. The purpose here was not to enable these women to 'earn a living' but to provide them with a worthy pastime that would allow them to become knowledgeable patrons, able to both commission and create new embroideries for the church themselves.

Our understanding of the early medieval world

This project not only gives us a greater appreciation of embroideries and how they contributed to life in the early medieval British Isles and Ireland, it also informs our understanding of larger themes relating to the early medieval world. This was a period of great change with numerous migrations, expanding diplomatic and trade networks, and conversion from belief in pagan deities to Christianity, all of which can be traced through embroidery. There was interconnectedness across the early medieval world, which can be seen particularly well in embroidery produced in Anglo-Saxon England. One such example was the demand and resulting movement of precious materials through religious, diplomatic and later Viking trade routes from Byzantium and beyond to Anglo-Saxon England. Once the silks, pearls and other precious materials had arrived in English workshops they were turned into embroidery that was itself sought after, both locally and on the Continent. The *Liber Pontificalis*, for example, lists embroidered items that were given to Popes by visiting Anglo-Saxon kings.[1] Thus we can trace the circular movement of raw materials and finished embroideries across early medieval Europe, giving us an insight into larger patterns of gift exchange and trade, and the resulting economic development of different regions.

Embroidery can also shed light on the transfer of technologies, conquest and power, and the transference of ideas. The late 9th-centry Ingleby fragment is especially helpful in demonstrating the movement of technology. This piece uses looped stitch, a very

---

1 R. Davis, trans., *The Lives of the Ninth-Century Popes (Liber Pontificalis)* (Liverpool: Liverpool University Press, 1995), 186–187.

old stitch form, and is worked in pulled silver metal thread. Crafts people working at the trading settlement of Birka in Sweden appear to have learnt how to produce this form of metal thread from Sami people who reside in northern Sweden. When the Vikings raided and settled in early medieval England, they brought this new technology with them, thus meshing Sami knowledge with Viking and then Anglo-Saxon culture.

Power and international ambitions were demonstrated through embroideries commissioned to tell specific stories, such as the 11th-century Bayeux Tapestry. This large pictorial hanging depicts the power of the Normans and their ambitions for conquest. These themes were also expressed through commissioned embroideries the Norman elite sent back to Normandy. These pieces demonstrated the control and power the Norman elite now had over the Anglo-Saxons, spreading positive messages about the idea of expansion. Similar themes were also seen in embroideries commissioned by the Norman elite of Sicily in the 12th century.[2] In these examples a unified message of conquest and power was stitched into narrative or sumptuous embroideries. These travelled across early medieval Europe to be read and understood by those who saw them.

The movement of people meant that their cultural ideals also travelled. Once we learn how to 'read' the embroideries and their biographies, we can unpick the cultural and religious footprints incorporated into them. The 7th-century Kempston fragment is a prime example. The fine wool ground fabric and embroidery threads point to materials sourced and processed in Anglo-Saxon England. The design, an entwined knot motif, hints at either a pagan or early Anglo-Saxon form of Christian decoration. Its metaphorical meanings can be traced back to the Germanic world, giving us access to the early medieval mindset and pointing to cross-cultural ties. The monk Boniface's disdain for this design exemplifies the wider Church's viewpoint. The fragmentary embroidery was kept in a copper-alloy box which was buried with its owner. Similar traditions were followed in Scandinavian countries, linking the woman who owned the embroidery to those geographical areas and belief systems. From such a small fragment, measuring only 24 mm × 54 mm, we are able read the interwoven stories of people from across the early medieval world.

When we 'read' early medieval embroidery, as well as understand its technical attributes, we are able to unpick much larger ideas that spanned the early medieval world. By doing so we can categorically state that embroidery produced in the early medieval British Isles and Ireland was not simply a craft worked by elite women; it was an art form as highly regarded as metalwork, sculpture and manuscript illumination. It was accessible across society in a myriad of forms with a multitude of meanings, and it was created by women working at the height of their powers, whom society regarded with the highest respect. With their fame came status and demand for their work across early medieval Europe. Through their surviving embroideries, we can today stitch together their stories.

---

2 See C. Vernon, 'Dressing for Succession in Norman Italy: the mantle of King Roger II', *Al-Masāq Journal of the Medieval Mediterranean* (2018), DOI: https://www.tandfonline.com/doi/full/10.1080/09503110.2018.1551699.

Appendix 1

Catalogue

General note

The catalogue is in alphabetical order of embroidery, with embroideries named by find site location. The code for each embroidery in Table 1A–G appears in parenthesis. Where more than one embroidery was found at a single site, the pieces have been placed in chronological order and given an alphabetical designation, and an Introduction provides an overview of the group. For reasons of brevity, scholars who have worked on the pieces are referred to by surname only, with short form title and date supplied at end of the entries (full references supplied in bibliography). Cross-references to other relevant entries in the catalogue are given at the end of entries.

Alfriston (Alf)

Object Type: Impression of fabric with embroidery
Date: 5th–6th century(?)
Find Date: 1912
Measurements: none available
Embroidery Stitches: Unidentified
Find Location: Grave 20, Alfriston cemetery, Sussex
Present Location: Lost/no record – last known location Barbican House Museum, Lewes
Record Number: not supplied
Description: Possible embroidery preserved as an impression in rust on the inner side of the 'tabs' of an iron buckle (Fig. 6).
Compare: N/A
Sources: Griffith and Salzmann, 'An Anglo-Saxon Cemetery at Alfriston' (1914)

Bayeux (Bay)

Object Type: Embroidered wall hanging
Date: 11th century
Find Date: 1476: first documented in the Bayeux Cathedral inventory
Measurements: *c.* length: 68.38 m × height: 0.5 m (measurements taken over the centuries have varied due to different conditions under which measurements have taken place)
Embroidery Stitches: Chain stitch, laid-work, split stitch, stem stitch
Find Location: Bayeux Cathedral, Bayeux, France
Present Location: Musée de la Tapisserie de Bayeux
Record Number: Nº1, *manuscript brodé*
Description: A near complete long, narrow hanging embroidered (Pl. 5a) on a tabby weave ground fabric of bleached linen with an average of 22 Z-twisted threads per cm along the warp and 18 Z-twisted threads per cm along the weft. The surviving ground fabric is constructed from nine pieces of linen of varying lengths, which were sewn together using the open seam method. The embroidery is worked mainly in 2-ply S-twisted wool thread; there are a number of small areas also stitched in an unbleached linen thread. The wool was dyed in ten colours: pinkish/orange-red, brownish-violet red, mustard yellow, beige, blue-black, dark blue, mid-blue, dark green, mid-green and light green, and worked in four stitches: stem stitch, laid-work, split stitch and chain stitch in various combinations. Laid-work and stem stitch are the two most utilised stitches: the first for filling areas and the second for lines and outlines.

There have been a number of restoration and conservation interventions which have resulted in some areas of the ground fabric being patched with pieces of a newer linen textile. Bédat and Girault-Kurtzeman distinguish nine different categories of patch intervention over the hanging's life span. During the 19th century some of the embroidery was also restored. From the front these areas tend to be more garish than the original because the replacement wool was dyed with chemical dyes. Some of these areas have faded at different rates to the original work, leading to further variations in the colours. The back of the hanging can be seen on photographs taken during the 1982–1983 conservation programme and this makes clearer the areas of restoration because the untidy stitching with a lot of thread wastage is substandard when compared to the original.

Most scholars agree that much of the inspiration for the embroidery's design comes from manuscripts held in the monastery library at Canterbury or other art forms worked in the Canterbury Style. As a result, most researchers agree that the Bayeux Tapestry was made in Anglo-Saxon England.
Compare: Sutton Hoo B (wool thread worked on linen), Durham F (laid-work)
Sources: Bédat and Girault-Kurtzeman, 'Technical Study of the Bayeux Embroidery' (2004); Lester-Makin, 'The Front Tells the Story' (2016); Lester-Makin, 'Les six châteaux de la Tapisserie de Bayeux' (2018). For further details of the vast amount of published scholarship on the Bayeux Tapestry, see Brown, *The Bayeux Tapestry, a sourcebook* (2013)

Coventry (Cov)

Object Type: Child's leather turnshoe, right foot; incomplete
Date: Late 11th–early 12th century
Find Date: 1927–1949 (Mr J.B. Shelton excavated a number of sites around Coventry, including Bull Inn, between 1927 and 1949 but left no records with the museum)
Measurements: Embroidery length: 45 mm × width: 5 mm–7 mm
Embroidery Stitches: Raised plait
Find Location: Bull Inn, Coventry
Present Location: Herbert Art Gallery, Coventry
Record Number: 49/185/12
Description: The majority of the upper right side of the shoe survives (Pl. 6b). The vamp is worked from one piece of leather and has a single embroidered stripe running from ankle to toe. The shoe was constructed in the one-piece economy style and had an insert missing from the inside quarter; the toe end formed a blunt point and the top-edge had been worked with a very fine binding that formed a scalloped edge. There are also two small slots cut into the upper part of each quarter so that a thong could be passed through and around the ankle to hold the shoe in place. The wool embroidery survives fairly intact for just under half of its original length. Thomas thought that it had been worked in two parallel lines in either a tight over stitch or cross stitch.
Compare: London B, London C, Winchester (embroidered vamp)
Sources: Thomas, *Medieval Footwear* (1980)

Dublin (Dub)

Object Type: Fragment of embroidery with looped stitch
Date: *c*. late 10th–early 11th century
Find Date: During 1974–1981 excavations
Measurements: none available
Embroidery Stitches: Looped stitch
Find Location: Wood Quay/Fishamble Street, Dublin
Present Location: National Museum of Ireland
Record Number: not supplied
Description: Fragment with looped stitch found during excavation. The find is still being analysed and information is sparse. Frances Pritchard (who is undertaking analysis of the embroidery: report forthcoming), has confirmed that the looped stitch appears to cover a seam.[1]
Compare: Ingleby, Orkney, Sutton Hoo A, York B (looped stitch)

---

1 Pers. comm. Frances Pritchard (2019).

Durham

Introduction

These seven Durham embroideries are all kept at Durham Conservation Centre when they are not on display in the Cathedral.

Durham A–E, known as the Cuthbert embroideries, were discovered in the tomb of St Cuthbert at Durham Cathedral when it was opened by Reverend J. Raine in 1827.

The Cuthbert embroideries are principally known for the three items embroidered with gold and silk on silk ground fabric, classified here as Durham E, D and C, consisting, respectively, of a stole, a maniple, and an item known as 'maniple II' or 'the small maniple', which Coatsworth has influentially argued is more likely to have been a pair of ribbons stitched together at some point. Durham A and B are less spectacular fragments of two separate bands.

At an unknown date a piece of the stole (E) and a section of the soumak band (A) were taken from the Cathedral and, in the early 19th century, given to Ushaw College, Durham, by Professor John Lingard, a historian based at that college.[2] These pieces were eventually relocated to the Durham University Conservation Centre for conservation and storage. Since the piece of stole is unequivocally identifiable as part of Durham E it does not have a separate catalogue entry although it is physically a discrete entity.

Sources for A–E: Bailey, 'St Cuthbert's Relics: some neglected evidence' (2002); Baldwin Brown and Christie, 'S. Cuthbert's Stole and Maniple' (April 1913/May 1913); Christie, *English Medieval Embroidery* (1938); Coatsworth, 'Embroideries from Tomb of Cuthbert' (2001); Crowfoot, 'The Braids' (1956); Granger-Taylor, 'The Weft-Patterned Silks' (2002); Plenderleith, 'The Stole and Maniples (a) the technique' (1956)

Durham F and G are an entirely separate set of embroideries found in 1795 in the tomb of Bishop William of St Calais, also located in Durham Cathedral. These have had less of scholarly attention.

Sources for F–G: Coatsworth, 'Stitches in Time' (2005); Ivy, *Embroideries at Durham* (1997)

Durham A (Dur A)

Object Type: Tablet-woven soumak band with possible embroidery; six fragments
Date: *c.* late 8th(?) century onwards
Find Date: 1827
Measurements: Width of band: 33 mm; Possible embroidery: 2 mm; Length: left band (possible embroidery and plait): 57 mm–58 mm; right band (possible embroidery and plait): 71 mm
Embroidery Stitches: Satin stitch(?)
Find Location: St Cuthbert's tomb, Durham Cathedral, County Durham

---

2 Pers. comm. Claire Marsland, Ushaw College, Durham (2012).

Present Location: Stored at the Durham University Conservation Centre, County Durham
Record Number: not supplied
Description: Six fragments of tablet-woven soumak band mounted on conservation fabric and framed. Pl. 8a shows a piece that was, at some point, separated from these six fragments and given to Ushaw College. The bands were originally examined in the 1930s by Crowfoot. Granger-Taylor re-analysed the band in the 1980s. The band was originally attached to two fabrics of silk, a red one of tabby weave and a twill weave dyed reddish-brown. Crowfoot hypothesised that the silks were already set out in their final form when the soumak band was attached to them because the surviving red silk thread used to sew the band to the silks passed through both silk backings as well as the band.

J.F. Flanagan described how the soumak was woven into a repeating five-branched palm leaf and rose-petal motif. The pattern was worked in six colours, red, brownish-red, light brown, purple, blue and gold. Crowfoot suggested that there may originally have been more shades that have since faded and described how the embroidery/decorative stitch work was worked in at least the five colours of silk thread, blue, brown, purple, red and gold. These were sewn in blocks of differing lengths that may have created a pattern in itself (Pl. 12b, c). Granger-Taylor thought that the blocks may have represented an embroidered version of the *latté* effect seen on Central Asian silks.

Microscopic images show that the decorative stitch work is extremely fine and each stitch is packed close to the previous one (Pl. 12b, c) indicating that the stitch work is not buttonhole stitch (as previously hypothesised) but a wrapped stitch, much like satin stitch.
Compare: Durham B, Mitchell's Hill, Orkney, Utrecht (embroidery on tablet-woven band)

Durham B (Dur B)

Object Type: Tablet-woven band in tabby weave with embroidery on either side of the central band
Date: c. late 8th(?) century onwards
Find Date: 1827
Measurements: Length: 25 mm × width: 20 mm (at widest surviving point)
Embroidery Stitches: Buttonhole stitch(?), satin stitch(?)
Find Location: St Cuthbert's tomb, Durham Cathedral, County Durham
Present Location: Stored at the Durham University Conservation Centre, County Durham
Record Number: not supplied
Description: Tablet-woven band in tabby weave with embroidery on either side of the central band mounted on conservation net and framed.

Grace Crowfoot and Hero Granger-Taylor both noted that the decorative stitching is the same as that on Durham A, one of a number of similarities. Granger-Taylor also noted that from the paired tablet twists outwards (that is, the outer edge of the central band to the band's edge), these pieces are identical. The brocaded decoration on Durham B, and the thread used for the soumak weave on Durham A, and the decorative stitch work, are all the same. The Durham A band was also found attached to an amber silk similar to the one onto which the Durham B band was sewn. As a result, Crowfoot thought both bands had come from the same item; Granger-Taylor was more cautious, suggesting that the difference in size may point towards B belonging to another contemporary garment.

Granger-Taylor observed that there were originally two coloured silk threads, which she judged to be red and yellow. She was also able to determine that there was brocaded decoration in both the warp and weft of the tablet-weaving, forming a diamond pattern of the two colours, resulting in a chequered effect. She also refers to blue thread visible amongst the surviving embroidery, again suggesting affinities to Durham A in use of colour. This, and aspects of Crowfoot's hypothesis concerning its construction raise the possibility that the same individuals in a single workshop made both bands.

Compare: Durham A, Mitchell's Hill, Orkney, Utrecht (embroidery on tablet-woven band)

Durham C (Dur C)

Object Type: Embroidered band: gold and silk on silk ground
Date: *c.* 909–916
Find Date: 1827
Measurements: Length: 66 mm × width: 49 mm (in present layout)
Embroidery Stitches: Couched work, stem stitch
Find Location: St Cuthbert's tomb, Durham Cathedral, County Durham
Present Location: From 2018, part of new display in Durham Cathedral; when not on display stored in Durham University Conservation Centre
Record Number: not supplied
Description: Embroidered band, gold and silk threads on silk ground fabrics. Known by scholarly convention as 'Maniple II' or the small maniple. Durham's new display identifies it as a girdle (Pl. 4a, b).

What we see today is two narrow bands of embroidery that have been sewn together length-ways to create a wider piece. One end is wedge-shaped. Viewed from the front, the foliage design is embroidered in red silk thread with a background couched in gold thread. From the back, the foliage design is worked in couched gold thread and the background in red silk thread. The piece has a reversible design of a central foliage stem with lateral leaves, flowers and fruit inhabited by small birds and animals.

There is evidence that this piece started life as a pair of narrow ribbons suitable for a head-dress or cloak. The embroidered bands were most likely made in two halves with wedge-shaped tabs attached to the lower ends, and at some point in the past the two halves were sewn together side-by-side, along the length. Although the evidence is unclear, it is probable that the embroideries were already part of the single piece we see today when they were found in the coffin of St Cuthbert in 1827. The reasons for the remodelling are not clear. Plenderleith suggested that the piece was originally two separate embroideries joined together at what is now the top end, creating a girdle measuring approximately 24 mm wide × 1272 mm long. She thought the change may have been made because the embroidered strip twisted and distorted over time.

However, Coatsworth noted that if the two bands had originally been joined together by the top ends, the design would not have flowed evenly from one band to the other, suggesting it was not designed in this way. She suggested that the two bands were never joined together end to end and were meant to be separate, to be used as two ribbons that could be attached to either side of a headdress or a cloak. The fact that the bands were embroidered on both sides indicates that they were supposed to be viewed front and back as they would have been if swinging loose as ribbons. She cited examples of illustrations where both men and women wear ribbons attached to headdresses or cloaks, most of which end with wedge-shaped tabs. Note also that if the two bands were attached top end to top end and used as a girdle, the design would flow the correct way on one band, and upside down on the other. The design only lies the correct way on both bands when they are seen side-by-side, as in their present mounting.

Another reason for arguing against the girdle and for the ribbons is that wrapped round the wearer's body at the waist, the embroidery on the reverse would be unseen and liable to wear and damage, whereas ribbons would not get damaged in the same way. Plenderleith noted that the top ends of the bands had been finished off properly rather than cut. If the bands were originally ribbons they would have been joined to another textile at this point. This would account for the partially repeated pattern that appears at the top end of each band, which looks planned. Coatsworth pointed out that if the bands had been joined as a girdle there must be a section missing.

Plenderleith discovered that in addition to the original net on which the embroidery was worked, there were two linings. The original net and the early lining are likely to be the same as those used in the construction of the stole and maniple (Durham D and E). The second lining is formed of the same fabric as a second lining found on the stole and maniple and was probably attached to all three pieces under the instruction of Reverend Dr Raine. The similarity of the original ground fabric and lining also suggests that the three embroideries were made in the same workshop and within the same time period. The stem stitch on this piece is in red silk of the same thickness as the red silk stem stitch outlines on the stole and maniple. This suggests batch materials being acquired specifically for particular sections of the work. However, the evidence does not necessarily demonstrate that the three pieces were a set, just that they were made with materials from the same batch. An ecclesiastical role for all three objects cannot be taken for granted: the secular dress accessory of ribbons is a plausible function.

Although the embroidery is reversible, the front and back were not worked on the same ground fabric at the same time. The front was embroidered on one piece of netting and the back on another, and the two finished pieces must then have been sewn wrong sides together. The front band was embroidered with a couched gold thread background and a stem stitch filling of red silk thread. The embroidery on the back has been reversed, so the background is worked in stem stitch of red silk thread and the design is couched in gold.

For discussion of Durham C's role as relics in Cuthbert's tomb, occurring when they became caught up in royal politics of the 10th century see Durham D, below.
Compare: Durham D, Durham E, Milan (style)

Durham D (Dur D)
Object Type: Stole with embroidery; partially complete
Date: c. 910–916
Find Date: 1827
Measurements: Length: 1.94 m (total length of seven remaining fragments as mounted); Terminal panels: length: 60 mm × width: 60 mm; Ushaw fragment: length: 33 mm × width: 18 mm of which embroidery is: length: 33 mm × width: 10 mm
Embroidery Stitches: Couched work, split stitch, stem stitch
Find Location: St Cuthbert's tomb, Durham Cathedral, County Durham
Present Location: From 2018, part of new display in Durham Cathedral; when not on display stored in Durham University Conservation Centre
Record Number: not supplied
Description: Partially complete stole (a scarf-like vestment worn over the shoulders with the ends crossed over and hanging almost to floor-length (see Glossary)) (Pl. 3a), with embroidery, gold and silk threads on silk ground fabrics, consists of embroidered images of 16 full length prophets, one on either side of an *Agnus Dei* symbol set in a quatrefoil, and the head of St Thomas and St James on the front of the end tabs with the name of the commissioner and receiver on the reverse.

The stole was originally made up of two lengths of embroidery joined at the centre with a quatrefoil-shaped panel stitched with the *Agnus Dei*. This panel would have sat at the back of the neck of the celebrant as he wore the stole. The two lengths of the stole were divided into sections that each contained full-length figures of the sixteen Major and Minor Prophets. Remnant of a similar design can be seen on the Worcester fragments. The end terminals each display an embroidered half-figure of a saint on the front and an inscription naming the original donor and intended recipient on the reverse.

The embroidered inscription, which also appears on the terminals of the maniple, states, ÆLFFLÆD FIERI PRECEPIT, and, PIO EPISCOPO FRIDESTANO. This information tells the viewer that a woman named Ælfflæd, assumed to be Queen Ælfflæd (who died before 916), commissioned the stole to be made for Bishop Frithestan, who was

Bishop of Winchester between 909 and 931. Queen Ælfflæd was King Edward the Elder's second wife. Edward the Elder's heir, Æthelstan, had been born during his first marriage and during the early years of his reign, Æthelstan's relationship with Bishop Frithestan was strained,[3] which may have led to the Bishop not receiving his gift. Æthelstan is recorded as giving a set of embroideries – a stole, maniple and girdle – to the community at Durham when he was on his way north to fight the Picts in c. 934,[4] and it would seem likely that at least Durham C to E are this gift.

The embroidery was worked in two types of silk thread, non-twisted and a thicker slightly S-twisted version, and couched gold thread. The thicker silk thread was used to stitch the stem stitch outlines and letters. The Ushaw fragment shows that stitches measure just 1 mm in length across the surviving letters (Pl. 22b). The couching was worked in a thread of gold foil wrapped round a red silk core. The method of working is the same as for Durham C, including the interlocking of the couching thread on the reverse.

Unfortunately the 19th-century lining was glued in place for both the stole and maniple. As the glue has dried and shrunk over time, it has pulled and twisted the vestments causing them to become somewhat distorted and brittle. Although close examination has shown that the gold thread is twisting and springing up from the ground fabric in places, the embroideries have survived in fairly good condition.

Compare: Durham C, Durham E, Milan (style), Worcester (stole)

Durham E (Dur E)

Object Type: Maniple with embroidery; almost complete
Date: c. 909–916
Find Date: 1827
Measurements: Length, including terminals: 806 mm × width, including band: 60 mm; Band: 8 mm; Terminal panels: 60 mm × 60 mm
Embroidery Stitches: Couched work, split stitch, stem stitch
Find Location: St Cuthbert's tomb, Durham Cathedral, County Durham
Present Location: From 2018 part of new display in Durham Cathedral; when not on display stored in Durham University Conservation Centre
Record Number: none supplied

---

3   M. Gretsch, 'The Junius Psalter Gloss', in *Edward the Elder 899-924*, eds N.J. Higham and D.H. Hill (London: Routledge, 2001), 280–291 (290); S. Keynes, ed., *The Liber Vitae of the New Minster and Hyde Abbey Winchester, British Library Stowe 944* (Copenhagen: Rosenkilde and Bagger, 1996), 19–21.
4   T. Arnold, ed., 'Historia de sancto Cuthberto', in *Symeonis monachi Opera Omnia*, Rolls Series, 1 (London: Longman, 1882), 196–214; D. Rollason, trans. and ed., *Libellus de Exordio Atque Procursu Istius, hoc est Dunhelmensis, Ecclesie: tract on the origins and progress of this the church of Durham* (Oxford: Clarendon: 2000), 134–137.

Description: The almost complete embroidered maniple, a scarf-like vestment hung over the wrist, with a fragment missing from one end, is worked in gold and silk threads on silk ground fabrics (Pl. 3b). In the centre of the design, which would lie over the wearer's arm, is a quatrefoil-shaped panel embroidered with 'The Hand of God' and the inscription, *Dextera Dei*. There are full length figures of a Pope and his Deacon located on either side, each standing on a mound with a canopy of foliage at the top, and the heads of St John the Baptist and St John the Evangelist on the front of the end tabs with the names of the commissioner and receiver on the reverse (Pl. 18e).

The maniple was constructed using the same working techniques and stitches as the stole. Plenderleith found that although the materials are essentially the same, the maniple had been embroidered even more finely: 51 gold threads couched to every centimetre on a ground fabric of 25 × 25 threads per cm, as opposed to 44 gold threads per cm on a foundation fabric of 22 × 22 threads per cm.

It is obvious from the design, workmanship and inscriptions that the stole and maniple were intended as a set and produced in the same workshop.

Compare: Durham C, Durham D, Milan (style), Worcester (maniple)

Durham F (Dur F)

Object Type: Fabric with embroidery
Date: 1060–1096
Find Date: 1795
Measurements: Entire fabric: width: 19 mm–83 mm × height: 30 mm–74 mm; Embroidery: width: 3.5 mm–5.5 mm × length: 9 mm–67 mm
Embroidery Stitches: Couched work, laid-work, satin stitch, stem stitch
Find Location: The grave of William of St Calais, Durham Cathedral, County Durham
Present Location: Stored at the Durham University Conservation Centre
Record Number: none supplied
Description: This object is one of two small pieces of embroidery discovered in the tomb of Bishop William of St Calais, who was bishop of Durham between 1081–1096. The embroidery is worked on a silk ground fabric using silk and silver-gilt threads (Pl. 8b). The design consists of interweaving foliage that sprouts from a stylised base. The embroidery probably came from an ecclesiastical vestment.

This is the only surviving fragment from this early period to contain laid-work in silk threads; the Bayeux Tapestry involves the same stitch worked in wool. Jill Ivy noted that the use of silver-gilt thread is usually found on post-Conquest embroideries; the same is the case with regard to laid-work. As such, this fragment is one of four surviving pieces in England, alongside Bayeux, Durham G and Worcester, that can be described as transitional embroideries; that is, they show the evolution of embroidery techniques from Anglo-Saxon embroidery to the later *Opus Anglicanum*.

Compare: Bayeux (laid-work), Durham G, Worcester (silver-gilt and silk threads)

Durham G (Dur G)

Object Type: Fabric with embroidery
Date: 1060–1096
Find Date: 1795
Measurements: Entire fabric: width: 17 mm–111 mm × length: 20 mm–86 mm; Embroidery: width: 16 mm–92 mm × length: 36 mm–71 mm
Embroidery Stitches: Stem stitch, underside couching
Find Location: The grave of William of St Calais, Durham Cathedral, County Durham
Present Location: Stored at the Durham University Conservation Centre
Record Number: none supplied
Description: This piece is the second fragment of embroidery found in the tomb of Bishop William of St Calais. As the photograph indicates, this piece has been cut down at some point: note the sharp cut lines around the top, left side and base of the fragment (Pl. 9a). The shape of the fragment is tantalising: curved along the bottom edge, with a straight hem to the right; but what it represents is not clear. The ground fabric seems originally to have extended further but it is difficult to deduce the original layout.

The embroidery is worked in under-side couching in silver-gilt threads on a silk ground fabric. The design incorporates griffins and lions encased in roundels with sprigs of foliage projecting from them. The workmanship is of a high standard. The threads are spaced regularly and close together, and have been underside-couched with precision. The embroidery has been underside couched over single threads using the alternating brick pattern. The design was a popular one: the arcade and roundel strips now housed in Maaseik from the late 8th to 9th centuries have similar designs.
Compare: Durham F, Worcester (silver-gilt and silk threads), Maaseik A and Maaseik B (design), Worcester (underside couching)

Ingleby (Ing)

Object Type: Carbonised metal thread embroidery with some carbonised fabric; two fragments
Date: *c.* 878–900
Find Date: 1955
Measurements: Fragment one: length: 11 mm × width: 6 mm–7 mm × depth: 9 mm; Fragment two: length: 17 mm × width: 8 mm × depth: 9 mm
Embroidery Stitches: Looped stitch
Find Location: Mound 11 (cremation burial), Heath Wood, near Ingleby, Derbyshire
Present Location: Derby Museum and Art Gallery
Record Number: <1985-225/16>
Description: Metal-work embroidery constructed from silver, now oxidised (Pl. 7a, b).

The embroidery was originally one piece that has split into two; the pieces still fit together extremely well, the top of fragment two slotting into to the bottom of

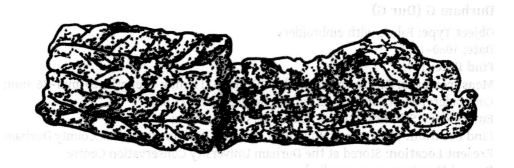

Figure 50. Line drawing of the Ingleby fragments as they would have originally been joined together, © Alexandra Lester-Makin, after Chaloner (2004)

fragment one (Fig. 50). The measurements given above were taken during a research visit to analyse the embroidery. In her report, Elisabeth Crowfoot had measured the pieces together totalling 28 mm long and 8 mm wide. The discrepancy between the width measurements can be accounted for by the loops at each end of every row which do not lie flat, bending to different degrees. The second fragment has also been warped and squashed together at one end, presumably during the cremation process. Crowfoot did not measure the embroidery's depth, which is consistent along both fragments.

Although the pieces look like woven metal, it has been determined that the pulled metal thread was sewn through a ground fabric, possibly silk or wool. When the embroideries were sent to the Shirley Institute (SI) tests found all but one sample reacted to tests as iron would. Testers at the SI suggested that the samples might be fibre bundles (see p. 42). Experiments undertaken for this research have also shown that wire thread can easily have been threaded through a large-eyed needle allowing it to be manipulated and passed through ground fabric as the embroidery was worked (Fig. 51). While the fragments were at SI the wire thread

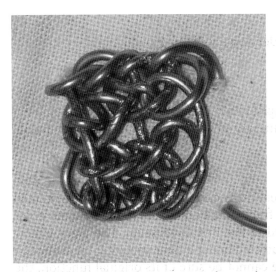

Figure 51. Looped stitch, experimental piece by author (26 × 30 mm), © Alexandra Lester-Makin

was also examined. It had a strong mauve colour suggesting silver metal. The wire was encrusted with sand but also had both reflective and opaque qualities leading to the suggestion that it had oxidised into chlorargyrite, horn silver.[5]

The embroidery of fragment one is constructed from 10 rows of looped stitch while fragment two has 12 surviving rows. As Crowfoot noted, both fragments contain four loops per row. In places the stitches at the ends of each row are slightly bent under, giving the impression of half stitches. Crowfoot also suggested that the embroidery was worked in a single length of thread. When one length of pulled metal thread was finished, a new length would have been started at the end of a row, probably with a knot placed to the reverse of the embroidery. The only indications of the new thread would be seen on the reverse of the work: the carrying thread that runs horizontally across the back of the work would be missing, being replaced by the starting knot. The top of fragment one has just such a knot on the reverse and at the top right of the section. It is therefore safe to suggest that this is the original start of this section for the embroidery.

The research experiments also showed that once the embroidery has been worked it is quite unyielding. Therefore a suggestion can be made concerning the missing wire thread that once passed through a now open stitch at the base of fragment two. It must have become weak, allowing the stiffened but brittle wire to snap in at least two places. The fragmented wire would then be able to work itself free leaving the open loop visible at the bottom of fragment two and separating it from the rest of the embroidery, now lost. This process must have occurred before the silver oxidised during the cremation because the chemical changes that took place during the burning of the metal solidified it into position.

Compare: Dublin, Orkney, Sutton Hoo A, York B (looped stitch)
Sources: Crowfoot, 'Appendix I. Objects found in Cremation Heath, Mound II' (1956)

---

5 Anon, Unpublished memorandum, English Heritage Archive. I am grateful to Claire Tsang and Kirsty Stonell Walker, Archive and Information Team at English Heritage, Fort Cumberland, for a copy of this memorandum.

Kempston (Kem)

Object Type: Fragment of wool embroidery on a woollen ground fabric
Date: 7th century
Find Date: 1863/1864 excavation
Measurements: 24 mm × 54 mm
Embroidery Stitches: Chain stitch(?), split stitch, stem stitch
Find Location: Probably grave 71, cemetery at a gravel quarry near Kempston, Bedfordshire
Present Location: British Museum
Record Number: <Bd1>(?), <484814001>
Description: A number of wool embroidery fragments originally discovered in a small copper alloy box (Pl. 2b, 12d). Box and fragments became separated at some point between the find and after they arrived at the British Museum. The piece is now mounted on conservation fabric and protected under a clear perpex cover. Although the remaining fragment is small and we cannot be sure of its original function, the design of entwined knots/beasts, possibly symbolising the World Serpent, suggests the decoration was meant as a form of protection for the person who wore or owned it. The fineness of the stitching indicates the embroidery came from what was originally an expensive piece which must have belonged to someone of status and/or wealth. That the fragment was retained after the original textile was no longer of functional use points to the original owner possibly being a prominent religious person. The embroidery appears to have been kept as a form of amulet/relic, either providing protection for the carrier or used as a touch relic because of its powerful association with an original owner (for extended discussion see Chapter 3).
Compare: N/A
Sources: Crowfoot, 'Textile Fragments' (1990)

Llangorse (Lla)

Object Type: Fabric with (possible) embroidery
Date: Late 9th–10th century
Find Date: During the 1990 season of the 1989–1993 excavation at Llangorse Lake
Measurements: Initial bundle measured 220 mm × 135 mm × 65 mm deep. Conservation established it consisted of over 40 layers of fabric which opened out to measure more than 8000 cm^2; with approx. 780 cm^2 decorated
Embroidery Stitches: Counted work, stem stitch
Find Location: Crannóg in Llangorse Lake near Brecon, south-east Wales
Present Location: The National Museum and Gallery, Cardiff
Record Number: archNum: 91.5H/[412]
Description: A large piece of charred textile discovered submerged at the bottom of Llangorse Lake on the man-made island, called a crannóg (probably the site of the royal residence of the early medieval rulers of Brycheiniog, which dendrochronological

dating showed had been built in the 890s and burnt down in the 10th century) (Fig. 2). It was found partially lying on charred wood submerged in waterlogged silts and the textile, charred wood and surrounding silts were lifted as one. The textile underwent careful cleaning and extensive conservation at the Department of Archaeology and Numismatics at the National Museum in Cardiff. Mumford believes the textile either fell gently or was lowered into the water, indicated by the way the folds in the textile have developed. Sediment between each one suggests that as the textile fell to the lake bed, it disturbed the silt, which got caught in the fabric as it concertinaed into its final resting position. The textile is held in a fridge to keep its condition stable.

The charring and blackening indicates the textile was caught up in a slow burning atmosphere and means only surface detail of the fibres can be studied with a scanning electron microscope (SEM) and Fourier transform infrared microscopy (FT-IR). Recent tests indicate that the textile was made from both silk and cellulose fibres. No trace of dyes have been found as yet. The decorative threads consist of two differently plied types of silk for background and design elements respectively.

Mumford, Granger-Taylor and Pritchard originally suggested the textile was a garment; possibly a shirt with a triangular insert forming part of a sleeve that was decorated with the embroidery. The evidence included a 45 mm-long belt loop made of silk, an 80 mm-wide hem and two eyelets buttonholed with silk thread. The eyelets appear functional, but it is not clear what their function is. Granger-Taylor and Pritchard's analysis also revealed the design of the garment must have been known before the decoration was applied as it was only found on areas that would likely have been seen, and on the insert. It also stops just before the seam joins, which are covered with finger-looped plaits sewn onto the textile as the seams were worked, and the buttonholed triangular reinforcements. But they pointed out that if the surviving textile was a sleeve, the insert would run from the armpit to wrist and decoration on this section would not be seen. They suggested it may have been more imaginatively used, referring to the so-called *Grande Robe*, a linen gown housed at Chelles, France, which has two half-triangular inserts starting on either side of the neckline and running down over the shoulders, and a triangular gore in the middle-back. In more recent publications, Mumford has suggested the textile could have been part of a tunic or dress.

High-resolution images were taken during conservation allowing the design to be drawn up. A decorative band with a main area bounded on two sides by a border incorporates stylised vine scrolls with leaves and bunches of grapes, and peacocks and ducks. The motifs are repeated and mirror imaged across the design. One border is decorated with a stylised geometric trefoil design and three-legged lions, with two spots and a raised tail. A less well preserved area of decoration located on the triangular insert includes birds in a border near two buttonholed eyelets and geometric shapes.

It is unclear whether the piece was tapestry-woven or embroidered owing to its fineness. Granger-Taylor concluded it was embroidery worked as counted stem stitch.

Binocular microscopy revealed that for every thread in the ground fabric there is one of embroidery (25 threads per cm in each case). Mumford believes mathematical precision is more suggestive of a woven technique such as soumak, and microscopic analysis has so far revealed no split or misplaced stitches. It is possible that it the design is a combination of woven technique and stitched work, although it is hard to imagine a rationale for this.
Compare: N/A
Sources: Campbell and Lane, 'Llangorse' (1989); Granger-Taylor and Pritchard, 'A Fine Quality Insular Embroidery from Llan-gors Crannóg' (2001); Mumford, 'The Conservation of the Llangorse Textile' (2002); Mumford, 'The Llangors Textile: an early medieval masterpiece' (2004); Mumford, Prosser and Taylor, 'Understanding an early medieval masterpiece' (2007); Redknap and Lane, 'The Early Medieval Crannog at Llangorse, Powys: an interim statement of the 1989–1993 seasons' (1994)

London

Introduction

The following three objects were excavated at Milk Street and the Guildhall in London. Due to the environmental conditions of the area (damp to waterlogged soils) a number of leather fragments, mainly from shoes, have survived, but most of the embroidery threads have not, so only three items include evidence that qualifies them for inclusion in this catalogue. Other items believed once to have been embellished with embroidery are listed in Table 3. Pieces of loose gold thread were also found at Guildhall and are included in Table 2.
Sources: Edwards, '302. MLK 76 [1053] <543>, Pit 55' (1991); MacConnoran, with Nailer, 'Complete Catalogue of Leather Items' (2007); Pritchard, 'Leather Work' (1991)

London A (Lon A)

Object Type: Leather top-band of shoe with embroidery
Date: 1050–1100
Find Date: 1976
Measurements: Surviving turned and decorated top-band: length: 198 mm × width: 30 mm
Embroidery Stitches: Running stitch(?), tunnel stitch(?)
Find Location: Pit 55, Milk Street, London City
Present Location: Museum of London Archive
Record Number: <MLK 76>, [1053], <543
Description: Fragmented right ankle-boot with a partially surviving top-band that was decoratively stitched (Pl. 5b). The upper survives to just below where a draw string would be located. It has a seam for a V-back sole and eight thong slots with surviving thongs are situated on the outside quarter. The top edge, side flap and

Figure 52. London A: detail of the embroidered top band, S-plied sewing thread, front view (×400), © Museum of London

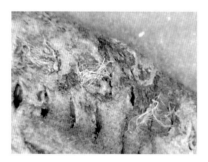

Figure 53. London A: detail showing S-plied sewing thread (×400), © Museum of London

Figure 54. London A: detail showing the folded leather being held in place by the top line of stitching, © Museum of London

opposite edge have stitch holes. Those on the top edge and opposite edge were not sewn. The V-back sole also survives in a very worn condition. It has tunnel stitching on the lasting margin with 2-ply sewing thread of wool surviving in both the upper and the sole.

The top-band has been partially preserved with one end torn. The band was turned under and one row of incisions was made through both pieces of leather. A second line of slits was created below the first but only penetrated the outer layer of leather (Fig. 52). Along the bottom edge of the top-band is a row of slits that would have been used to sew the top-band to the upper of the ankle-boot. Microscopic imagery (Fig. 53) shows the remains of S-ply sewing thread in some of the slits. Pritchard states that fibre-expert Michael Ryder thought the thread was most likely wool. The surviving evidence suggests the two rows were decorated with a running stitch forming two parallel lines. The top line held the folded leather in place and was thus also functional (Fig. 54).

Compare: N/A

London B (Lon B)

Object Type: Leather shoe vamp with embroidery
Date: *c.* 1070s–1080s
Find Date: During the excavation period of the 1980s/1990s
Measurements: Shoe: length: 155 mm × width: 130 mm
Embroidery Stitches: Tunnel stitch
Find Location: Open Area 105, Guildhall, London City
Present Location: Museum of London Archaeology Archives
Record Number: <L66>, A[17548], A<4391>
Description: Vamp of an incomplete embroidered left shoe with a gently rounded toe-end.

The vamp fragment is torn and has been cut for re-use. The right side of the vamp has a

near-vertical edge seam and there is a butt seam at the edge of the throat, which is low cut. The vamp is decorated with four rows of fine stitch holes that do not penetrate the back (flesh side) of the leather. The four lines make up one vamp stripe. Within the holes there is some surviving thread that may be silk.
Compare: Coventry, London C, Winchester (embroidered vamp)

London C (Lon C)

Object Type: Leather shoe vamp with embroidery
Date: *c.* 1070s–1080s
Find Date: During the excavation period of the 1980s/1990s
Measurements: Shoe: length: 155 mm × width: 105 mm
Embroidery Stitches: Tunnel stitch(?)
Find Location: Open Area 105, Guildhall, London City
Present Location: Museum of London Archaeology Archives
Accession Find Number: <L67>, A[17548], A<4393>
Description: Incomplete embroidered left shoe vamp with a gently rounded toe-end. This piece has a lasting margin that has been evenly stitched. The vamp is fragmented and has been cut for re-use. There is a vamp stripe made up of four rows of fine stitch holes that possibly have some thread surviving in them. The shoe is very similar to London B.
Compare: Coventry, London B, Winchester (embroidered vamp)

Maaseik

Introduction

The Maaseik embroideries consist of eight separate pieces that are catalogued in three groups, by convention, because the design evidence suggests three autonomous or semi-autonomous groups. Maaseik A and B are each made up of two embroidered bands known by the designs worked on each as arcade and roundel strips respectively. The arcade strips depict rows of arches with foliage or animals; the roundel strips incorporate rows of circles containing animals. Maaseik C consists of four probable monograms (embroidered letters for application to other textiles).

    The eight embroideries were initially discovered applied to what is now called the *casula* (or chasuble) of saints Harlindis and Relindis, two sisters who founded an abbey church in Aldeneik, Belgium in the early 8th century (Pl. 6a, 10). However, the embroideries date to the late 8th to 9th centuries and were reconstituted as the composite *casula* at some later point in their history. According to Budny and Tweddle the *casula* was removed from the abbey church at Aldeneik to St Catherine's Church in Maaseik in 1571, where it lay forgotten until it was rediscovered in a reliquary in 1867. The embroideries were not subjected to detailed study until the mid-20th century.

Belgian textile specialist, Marguerite Calberg, published an article discussing the embroideries in 1951, dated them to the second half of the 9th century and classified them as of Anglo-Saxon origin. Between 1979 and the mid-1980s the embroideries were studied as part of a British Academy funded project resulting in two articles from Budny and Tweddle, who have convincingly argued that the embroideries were created in the same workshop and can be dated to late 700 to early 800 and that the embroideries form a set. In fact, although the technical evidence bears out the single workshop thesis, it is possible that that they were ordered by a single patron at different times or that more than one patron commissioned similar designs. The ground fabric used for each embroidery differs and the silk threads may have come from different batches, as discussed in Chapter 5. Moreover, since we do not know how the embroideries came to be at Aldeneik, it cannot be assumed they arrived at the same time. Thus a more prudent thesis is that the embroideries were made in a single workshop and may be meant to form a set but could have been created as separate pieces that eventually found their way to the same shrine. For discussion of design elements, see Chapter 4.

The items are now housed in the treasury at St Catherine's Church in Maaseik, in a specially constructed display case and laid out as individual pieces. The arcade and roundel strips of which A and B consist were sewn onto mounts covered with conservation fabric and all the embroideries have been laid onto a board also covered with conservation fabric. The mounting of the embroideries means study of the obverse is not possible.

Sources: Budny, 'Anglo-Saxon Embroideries' (1984); Budny and Tweddle, 'The Maaseik Embroideries' (1984); Budny and Tweddle, 'Early Medieval Textiles at Maaseik' (1985); Calberg, 'Tissus et Broderies Attribués aux Saintes Harlinde et Relinde' (1951); Coatsworth, 'Stitches in Time' (2005)

Maaseik A (Maa A)

Object Type: Two silk and gold embroidered bands
Date: Late 8th–early 9th century
Find Date: 2 September 1867
Measurements: Arcade strip 1: length: 630 mm × height: 95 mm; Arcade strip 2: length: 660 mm × height: 100 mm
Embroidery Stitches: Couched work, stem stitch
Find Location: Reliquary, St Catherine's Church, Maaseik, Belgium
Present Location: Display case in the treasury of St Catherine's Church, Maaseik, Belgium
Record Number: Arcade strip 1 & 2: < 15396>, < 1984.03010>
Description: Two long, thin bands of silk and gold embroidery worked in an inhabited arcade design populated by animals (Pl. 23). The bands are not complete, having been cut down at some point in their life (possibly when they were applied to the *casula*

to make them fit the space), and they differ slightly in size. On the *casula* each band lay on either side of the composite piece and each band-end would have butted up against one of the four monograms (Maaseik C), so the bands may have been cut down to allow the precise spaces needed for each monogram (Pl. 10).

The arcade strips show a lot of wear and tear (as do the other Maaseik pieces). It is difficult to postulate a thesis as to their use prior to donation to the shrine at Aldeneik in Belgium from the surviving evidence. After arrival at Aldeneik it is likely the pieces were recycled, and resulting in a change of role at least once after donation, and eventually in their re-use as part of the composite *casula*. Size indicates that both the arcade and roundel strips were most likely used as decorative bands on clothing or textiles from which they could be detached, prior to this. There is documentary evidence for the use of detachable bands on clothing: for example, in the *Liber Eliensis* Ealdorman Brythnoth is mentioned as having detached the edges from his cloak and donated them to the monastery at Ely when he was on his way to fight the Vikings in 991.[6]

Compare: Maaseik B

Maaseik B (Maa B)

Object Type: Two silk and gold embroidered bands
Date: Late 8th–early 9th century
Find Date: 2 September 1867
Measurements: Roundel strip 1: length: *c.* 190 mm × height: 187 mm; Roundel strip 2: length: *c.* 190 mm × height: 187 mm
Embroidery Stitches: Couched work, stem stitch
Find Location: Reliquary, St Catherine's Church, Maaseik, Belgium
Present Location: Display case in the treasury of St Catherine's Church, Maaseik, Belgium
Record Number: Roundel strip 1 & 2: <15396>, <1984.03010>
Description: Two rectangular bands of silk and gold embroidery worked in an inhabited roundel design populated by creatures (Pl. 19c). These two bands were found joined together to form a single panel lying across the middle of the *casula* (Pl. 10). When they were detached from it the band was found to be made of two separate sections. These were unpicked during the conservation process. The cut edges of each band were found to have been turned under when they were joined together. The join is not straight, following the uneven line of the cut edges. The band originally placed to the left was facing the correct way up, with the embroidered animals positioned upright, while the band to the right had been attached upside down. This indicates that while such embroideries were considered important and precious, their original design and function was discounted in re-use with their representational role on the

---

6 Blake, *Liber Eliensis*, 135; Fairweather, *Liber Eliensis*, 162.

casula taking precedence (as relics associated with the sister saints). See discussion of object biographies in Chapter 1.
Compare: Maaseik A; Oseberg E (design)

Maaseik C (Maa C)

Object Type: Four silk embroidered monograms
Date: Late 8th–early 9th century
Find Date: 2 September 1867
Measurements: Monogram 1: length: *c.* 120 mm × height: 131 mm; Monogram 2: length: *c.* 120 mm × height: 131 mm; Monogram 3: length: *c.* 120 mm × height: 131 mm; Monogram 4: *c.* 120 mm × height: 131 mm
Embroidery Stitches: Couched work, stem stitch
Find Location: Reliquary, St Catherine's Church, Maaseik, Belgium
Present Location: Display case in the treasury of St Catherine's Church, Maaseik, Belgium
Record Number: Monogram 1, 2, 3 & 4: < 15396>, < 1984.03010>
Description: Four silk embroideries worked on a painted fabric and cut out to form probable monograms. They were found applied to four corners of the *casula* at each end of the arcade strips and directly on top of what is called the David silk and the half-silk (see Pl. 32). The monograms differ in size and the arcade strips seem to have been cut and fitted between them with some precision and forethought on the part of the worker.

Of all the Maaseik embroideries the monograms have survived the least well. They have been thought to represent either an M and an A or an M and an O, which may refer to letters from the Virgin Mary's name, Maria, or the alpha and omega, beginning and end. By the time the monograms were applied to the *casula*, it appears that their original meaning had either been lost or did not matter, since two were applied upside down (as per the roundel strips, and with similar implications – see above).

Prior to use on the *casula* it is possible that the monograms were used as detachable items for clothing in the same way as the arcade and roundel strips. It is difficult to imagine what type of garment the monograms might have adorned. It is perhaps more likely that they would have embellished some form of soft furnishing. Coatsworth has suggested that if they represent the alpha and omega or another religious symbol, they could have been placed in the corners of an altar-cloth or cover. In this case, the fact that two of the monograms could have been sewn onto the textile upside down would make sense, as draped over an altar or table they would have faced the correct way to the viewer. Such a hypothesis would support the idea that while the Maaseik pieces were all made in the same workshop, and possibly commissioned by one patron, they may not have been a set but could have been envisaged as performing two or three differing functions.
Compare: N/A

Milan (Mil)

Object Type: Gold-work embroidery with fabric
Date: 10th century
Find Date: 14th century: first documented as being kept in the Basilica Ambrosiana sacristy
Measurements: Main embroidery: length: 345 mm–347 mm; width: 29 mm–32 mm; Ground fabric: length: 412 mm × width: 38 mm
Embroidery Stitches: Couched work, stem stitch
Find Location: The reliquary of St Ambrose in the Basilica Ambrosiana, Milan, Italy
Present Location: Viewing drawers in the upper aisle in the Basilica Ambrosiana, Milan, Italy
Record Number: S6 (the largest piece of gold embroidery); S4 (small fragments of the same); S11(tablet-woven inscription); S9 (fabric of silk)
Description: Embroidered gold-work remounted onto a more recent textile (Pl. 9b, 21a, b).

The main surviving section of the embroidery, S6, was remounted onto a later ground fabric of black silk damask in 1863. There are also 13 loose fragments of gold-work which were placed next to S6 when all the pieces were set in a double-sided glass frame during the 1940s conservation programme documented by Ugo Monneret de Villard. Some of the original ground fabric was found among the 13 fragments when they were examined in the 1940s. This showed the original ground fabric to be a violet and black silk, the same as another surviving example, S4. This silk also had fragments of gold-work attached.

Although the origin and date of this embroidery is still debated, its history as part of another textile has been compiled by Granger-Taylor. The now re-mounted section of embroidery was once attached to what is now called 'dalmatic two'. The 13 loose fragments were probably part of this too. It is known that in the 11th century, dalmatic two was wrapped in a *pallium* and given as a gift, or enshrined as a relic of Saint Ambrose, by Archbishop Ariberto (1018–1045). The date is confirmed by a tablet-woven inscription (S11) attached to the *pallium* that reads, *Sub hoc pallio tegitur dalmatica sci. Ambrosii + sub quo eandem dalmaticam texit dommus heribertus archiepiscopus*, and which Granger-Taylor translates as, 'under this pallium is covered the dalmatic[s?] of St Ambrose. Under it Father Ariberto, Archbishop, covered the said dalmatic'. A second inscription is woven in white or yellow thread onto a dark blue Islamic silk (S3) and has a repeating Kufic script referring to Nasr ad-dawlah Abu Nasr, 'Glory and duration [long life?] emir as-Sayyid al-Nasr ad-agcdl Dawlah Abu Nasr, who [may?] Allah prolong his life' (de Villard translated by Granger-Taylor). Abu Nasr was the ruler of upper Mesopotamia between AD 1010 and 1061 and because of the form the woven title takes, de Villard asserted that the silk was created before he became ruler of the city of Diyarbakr in AD 1024 or 1025, meaning the silk could be dated to between AD 1010 and 1024. However, the textiles from which the dalmatic itself was made are considerably older. According to Granger-Taylor, the base silk (S9) appears to be a tunic of Late Roman origin. Onto this have been added, over time, other silks, linen and the embroideries, with at least one, a purple damask silk, being attached after the dalmatic had stopped being of practical use.

In the 11th century the dalmatic seems to have been patched and re-lined before it was wrapped in other textiles and given to Saint Ambrose. In 1608 the dalmatics were moved from their original housing, which inventory evidence from the 14th, 16th and 17th centuries suggests was the Sacristy, to a specially made reliquary chest also stored in the Sacristy, where they were kept until 1863. Visitors were able to view the textiles and pieces are known to have been removed because Monsignor Rossi, the Prevosto, complained of this in a letter dated to 1862. In 1863 he removed the dalmatics and it was at this time that the embroidery was re-mounted on the new ground fabric. The dalmatics were then put on display in a glass case with only pieces of textile dating to the 11th century visible. The case was placed above the altar of St Ambrose where it stayed until May 1940, when the textiles were split up and taken for conservation at Milan Cathedral and the conservation laboratory of the Vatican Library, according to de Villard. After this they were retained in Milan Cathedral for safe keeping during World War II, and then in 1949 the dalmatics were returned to the basilica and put on display in the church's museum.

Those scholars who have studied the embroideries (Alberto de Capitani d'Arzago in the earlier decades of the 20th-century, Crowfoot in the 1950s and Granger-Taylor in the 1980s) agreed that the embroidery was worked in gold thread constructed from a core of red silk thread with a fine strip of pure gold wrapped around it, couched in place with a thread of red silk and outlined with a thicker red silk thread. My observations show that a further row of couched gold thread (which outlines three edges of the embroidered frame that surrounds the main design) differs from the thread described in previous analyses, being wound round a core of yellow silk. It is wound less neatly than the original gold thread and there are areas that appear tarnished, which suggests it is silver-gilt, as opposed to pure gold. The threads would need to be tested with a scanning electron microscope to confirm this hypothesis, but the evidence points to another chapter in the embroidery's biography. This embroidery's construction is discussed in more detail in Chapter 4. The dalmatics are now being conserved by a German team. Professor Sabine Schrenk of the University of Bonn and the Cologne-based textile conservationist Ulrike Reichert are in the process of surveying, conserving and re-housing the textiles in modern protective glass frames. It is housed in a special drawer cabinet in the upper aisle of the basilica, still in the 1940s frame prior to remounting. The gold-work and the 13 fragments are held in place between two pieces of glass with strips of sticky cellophane tape.

Compare: Durham C, Durham D, Durham E (style)
Sources: d'Arzago, *Antichi Tessuti della Basilica Ambrosiana* (1941); Budny, 'The Anglo-Saxon Embroideries at Maaseik' (1984); Budny and Tweddle, 'The Maaseik

Embroideries' (1984); Crowfoot, 'Note on a Fragment from Basilica Ambrosiana' (1956); Granger-Taylor, 'Two Dalmatics' (1983); Granger-Taylor, Unpublished extract; de Villard, 'Una Iscrizione Marwanide' (1940)

Mitchell's Hill (Mit)

Object Type: Embroidered tablet-woven band
Date: 6th century
Find Date: 1909
Measurements: Tablet-woven band: 28 mm × 13 mm; Embroidery: unknown
Embroidery Stitches: Buttonhole stitch
Find Location: Mitchell's Hill, Icklingham, Suffolk
Present Location: Unknown; Last known location: Ashmolean Museum Oxford
Record Number: AN1909.487.[i] (wrist-clasp with textile); AN1909.487.a.1 (textile)
Description: Possible embroidery on a fragment of tablet-woven band found sandwiched between a textile and one of two sets of wrist-clasps, the discovery context of which are unknown (Fig. 8). When Crowfoot examined the textiles in 1952 both wrist-clasps were housed in the Ashmolean Museum in Oxford but they have since gone missing.

The tablet-woven band with embroidery was found to the back of the right clasp of one pair. The band had been well preserved between the metal of the clasp and the extremely fragmented twill fabric. The surviving tablet weave consisted of 26 twists. The edge of the band had been sewn with blanket stitch to stop it from fraying. Crowfoot noted that the possible embroidery was worked in a 2-ply yarn stitched across the tablet weaving in an irregular form. Fibres from the textiles and tablet-woven bands were sent to the Wool Industries Research Association in Leeds for analysis. All fibres were undyed wool, on the tablet-woven bands. She does not indicate whether the yarn used on the embroidery was analysed. The embroidery is difficult to see on surviving photographs but it looks as if the stitches might be a form of satin or split stitch, which were in common use throughout the early medieval period. If the stitching is embroidery it would not have been seen behind the clasp so it is likely that the design would have continued along the tablet-woven band onto an area that would be seen.
Compare: Oseberg E, Sutton Hoo A, Utrecht, York B (other forms of functionally decorative embroidery)
Sources: Crowfoot, 'Anglo-Saxon Tablet Weaving' (1952)

Orkney (Ork)

Object Type: Child's hood with embroidery
Date: c. 250–615
Find Date: 1867

Measurements: Hood: middle forehead to beneath chin: 253 mm; Point of crown to bottom of tablet-woven bands at the back: 448 mm; Looped stitch seam: length: 448 mm; Rows of chain stitch: approx. lengths: 35 mm, 96 mm, 69 mm, 86 mm
Embroidery Stitches: Chain stitch, looped stitch
Find Location: Peat bog, St Andrew's Parish, Orkney
Present Location: National Museum of Scotland, Edinburgh
Record Number: <X.NA3>
Description: Child's hood (Pl. 1a) found in the 19th century in a bog in the parish of St Andrew, on Orkney. As the object was un-stratified (found outside a dateable context), the date of its production was established through radiocarbon dating in the early 1990s at the University of Oxford. The results calibrated the date for the wool of the textile to between *c*. AD 250–615.

When Audrey Henshall first examined the hood in the early 1950s, she found that it was constructed from three pieces of separately woven, wool, 2/2, herringbone stripe twill cut from a larger piece, evidenced by the lack of selvedges. At the same time Dr Ryder examined the yarn and found it to be of nine types, all medium hairy. During conservation in 1981, a dye test found no dyes were now present; the brown tone being the natural fibre colour. It also emerged that at some point unrecorded repairs had been undertaken on a split in the textile. The cotton thread used for the repair was only invented in 1877.

At some point, the hood has been recycled. Jacqui Wood undertook a project earlier this century in which she analysed the hood and made a replica, discovering that it had been partially made from bands that had been previously cut and used on another garment.

The hood is constructed from fabric that has been cut and sewn to shape. The edges of the hood were turned over twice and stitched in place on the front side. The edges around the face of the wearer were also folded over twice and hemmed on what would become the visible side. The seams were then joined together using over-sew stitch. Two wedge-shaped pieces were added on either side of the hood at the face edge. Henshall thought they were used to mend areas of wear. At these points double rows of chain stitch worked in dark brown and yellow wool partially survive. Both Henshall and Marjory Findlay, the 1980s conservationist, believed these were decorative in part, as they do not actually hold the darning or patching in place. Findlay also noted the embroidery may have carried on past the applied fabric to meet at the back of the head.

Around the shoulder and back edge of the hood a narrow tablet-woven band measuring 20 mm wide has been applied. It lies over the textile with the upper edge sewn in place with what Henshall notes as regular stitches, which are covered with a complex form of looped stitch. Its lower edge was sewn to a second tablet-woven band measuring 70 mm wide. The narrow band covers the raw edge of the hood, which was turned and hemmed to the facing side. To this hemmed textile edge the second, wider tablet-woven band was attached. From this hung an integral fringe that measures 280 mm at the front and 303 mm at the rear. This band, the one that Wood noted as recycled, started at the back left of the shoulder edge. Although it was longer than the circumference of the hood edge, the maker decided not to cut it but rather to wrap it round the hood twice. This doubles the band and fringe along most of the hood, but there was a shortfall resulting in only a single layer over a 200 mm section at the back of the hood. Where the band lay at the sides of the face edge but below shoulder level, there are eight pairs of holes, some of which still contain leather thongs. A number of the thongs are still knotted and Thea Gabra-Sanders has suggested that they may have been used to attach the hood to another garment, or hold a draw string. Wood discovered that the hood was the right size for a child, with the 140 mm long seam that ran from the crown to the edge being too small to fit an adult head.

The example from Orkney is the most elaborate surviving version of looped stitch, using three lengths of thread simultaneously to create a doubled-ended interwoven loop that does not pass to the reverse of the fabric, but weaves through the stitches holding the seam together. That it uses three threads means that it could be worked in more than one colour, giving the stitch an additional dimension to its decorative quality, on top of the stitch itself. The version on the Orkney hood seems, however, to have been worked in one colour only.

Gabra-Sanders notes that hoods that extend down to cover the shoulders were known in Europe from the Roman period. By the 13th century they had evolved into a fashion statement for the aristocracy and were being used as an essential piece of clothing by those of lower rank. The Orkney example thus fits into a long tradition of hoods with other shoulder wear.

Compare: Ingleby, Sutton Hoo A, York B (looped stitch)
Sources: Findlay, 'Report on the Conservation of the Orkney Hood' (1984); Gabra-Sanders, 'Orkney Hood, Re-Dated and Re-Considered' (2001); Hedges *et al.*, 'Radiocarbon Dates' (1993); Henshall, 'Early Textiles Found in Scotland' (1951–2); Wood, 'Orkney Hood' (2003)

Oseberg

Introduction

The Oseberg Embroideries comprise 12 fragments of wool and silk that came from a richly furnished ship burial discovered on farmland in Norway in 1903 and excavated in 1904. The burial was hermetically sealed under a mound measuring 44 metres across and 6 metres high that had been constructed from closely packed pieces of turf. As a result, the preservation of the ship and contents, particularly the organic artefacts, was extremely good, even though it had been plundered at some point in its past, which would have interrupted the burial's protected climate. The wood of the ship and burial chamber have been dendrochronologically dated and the ship was built *c.* AD 800 while the burial chamber was dated to AD 834. Christensen and Nöcket suggest the ground fabric of the wool embroideries was woven in Ireland while the silk embroideries were created in Anglo-Saxon England. Photographs of the textiles and detailed line drawings of the stitch work included in the fourth volume of the excavation report are useful resources for researchers.

Sources: Bonde and Christensen, *Dendrokronologisk datering* (1991/1992); Christensen and Nöcket, *Osebergfunnet: Bind IV* (2006) (Vol. 4 of the Excavation Report); Coatsworth, 'Stitches in Time' (2005); Ingstad, 'Tekstilene i Osebergskipet' (1993); Nöckert, 'The Oseberg Textiles and the Orient' (2002)

Oseberg A (Ose A)

Object Type: Fragment of embroidery on fabric
Date: Late 8th–9th century
Find Date: 1904
Measurements: *c.* 30 mm × 30 mm
Embroidery Stitches: Couched work(?), stem stitch(?)
Find Location: Ship burial at Oseberg farm, Norway
Present Location: Viking Boat Museum, Oslo, Norway
Record Number: <12L1, Group N, number 16>
Description: A small fragment of fine tabby weave wool fabric, partially embroidered (Fig. 10). The textile was found in a bundle with numerous other fragments, of which three others are also decoratively stitched (B, C and D). This piece forms part of what the report authors call 'group N', and is red in colour. The origin of the ground fabric is thought to have been Ireland. This does not necessarily mean the piece was embroidered in Ireland, since textile trading occurred throughout the period and across the North Sea Zone.

Figure 10 indicates that there are remains of stem stitch embroidery and possibly a couched thread to the top left of the surviving stitch work. The stitches on this fragment vary between 2.5 and 5 mm in length, so although the stitching is quite fine it is not as painstakingly accomplished as other catalogue entries. The thread is wool but its dye, if there was any, has been lost.

In the excavation report catalogue, the embroidery is described as insignificant and it is true that the remaining sample is small and disjointed. No other technical data, such as the make-up of the thread, is supplied. However, as the Kempston fragment shows, small fragmentary pieces can yield valuable information. Oseberg finds demonstrate that embroidered textiles were being created and traded, taken as booty or given as gifts across the North Sea Zone, indicating that such decorative work was prized, even, as in this case, where the ground fabric and thread are not made of what would be considered precious materials like silk and gold.

Compare: Oseberg B, C and D

Figure 55. A strap-end from York, sf 7306, decorated with an entwinned knot and animal features. © York Archaeological Trust

Oseberg B (Ose B)

Object Type: Fragment with embroidery
Date: Late 8th–9th century
Find Date: 1904
Measurements: not available
Embroidery Stitches: Stem stitch(?)
Find Location: Ship burial at Oseberg farm, Norway
Present Location: Viking Boat Museum, Oslo, Norway
Record Number: Unclear, possibly <12L1, 18> Group N
Description: A fragment of wool cloth, with embroidery associated, found in the same bundle as Oseberg A. This piece is also red in colour. B forms part of one of two layers of tabby woven wool cloth. The fabric had a thread count of 22–26 × 13–16 threads per cm and was of excellent quality, closely and firmly woven. The piece also has some surviving soumak weave attached. As for fragment A, Ireland is given as the most likely place of origin for the ground fabric. Also associated with this textile, but not attached to it, are loose fragments of embroidery.

The stitching was worked in a wool yarn constructed from two Z-spun threads that were S-twisted together, with the loose threads made in the same manner. The line drawing associated with the excavation report's catalogue entry (Fig. 11) indicates that the workmanship is similar to that of Oseberg A. The stitching and cord arrangement appears to form a possible animal's head with two knots as the eyes. If this is correct, the style is reminiscent of strap ends (Fig. 55) and other forms of metalwork that date from throughout the early medieval period.
Compare: Oseberg A, C and D

Oseberg C (Ose C)

Object Type: Fragment with embroidery
Date: Late 8th–9th century
Find Date: 1904
Measurements: not available
Embroidery Stitches: Unidentified
Find Location: Ship burial at Oseberg farm, Norway
Present Location: Viking Boat Museum, Oslo, Norway
Record Number: Unclear, possibly <12L1, (2) 6> Group N
Description: Fragment of ground fabric made from finely spun wool with embroidery attached (Fig. 12).

The textile measures *c.* 30 mm by 10 mm. The report gives a thread count and quality of textile comparable to Oseberg B, and again posits Ireland as the country of origin. C is embroidered with five rings worked in two Z-spun wool threads S-plied together. One thread forms the core of which the circles were made, around which a second yarn, of the same fibre and construction, has been wound.

Depending on how the decorative circles were attached to the ground textile this piece could technically be a form of appliqué, not embroidery. Ingstad previously noted that similar appliqués and embroideries relating to clothing have been discovered at Birka. She went on to suggest that Oseberg C may have belonged to a sleeveless tunic worn over an undergarment. The tunic would have been gathered at the back while the front would have been decorated with appliqué and embroidery.

Also associated with this fragment (and entered under the same catalogue entry) are other fragments of decorative stitch work attached to a separate piece of fabric made either of wool or silk, which are mentioned only in passing in the excavation report, and by Ingstad.

Compare: Oseberg A, B and D

Oseberg D (Ose D)

Object Type: Small amounts of a fragment with embroidery
Date: Late 8th–9th century
Find Date: 1904
Measurements: 10 mm × 18 mm
Embroidery Stitches: Stem stitch(?)
Find Location: Ship burial at Oseberg farm, Norway
Present Location: Viking Boat Museum, Oslo, Norway
Record Number: <12L3, 3> Group N
Description: The catalogue entry in the excavation report states that the fragmented ground fabric is a tabby weave of wool, probably from Ireland, as for A/B/C. It is also red in colour. Some surviving soumak loops are woven into the textile next to the small amounts of extant embroidery. Discernible threads were made from yarn that was loosely Z-spun. They were shiny and originally dyed red and white, with the red yarn used to outline areas of the design stitched with white thread.

The combination of colours and technique of filling and outlining in different shades brings to mind the knot and entwined beast motifs on earlier embroideries such as the 7th-century Kempston fragment and the 8th-century Maaseik embroideries. The Oseberg ship itself has given its name to a style of Viking art using entwining motifs composed of a central section with outline. In fact variations of these motifs and constructional elements can be seen on work predating this era ranging from embroidery to metalwork, and from bone and ivory carving to stone sculpture in Anglo-Saxon England and across the Scandinavian world, where traditional northern myths and design ideas were incorporated across art forms. It is thus possible that an entwined motif of some description was embroidered on this piece. See also discussion in

Chapter 3. The embroidery was described by the report's authors as *lange kontursting*, long outline stitch, which is a form of stem stitch constructed from long individual stitches.
Compare: Oseberg A, B and C; Kempston, Maaseik A, Maaseik B (infill and outline technique)

Oseberg E (Ose E)

Object Type: Fragment with embroidery
Date: Late 8th–9th century
Find Date: 1904
Measurements: Width: 56 mm × height: 30 mm
Embroidery Stitches: Buttonhole stitch, couched work, satin stitch, split stitch, stem stitch
Find Location: Ship burial at Oseberg farm, Norway
Present Location: Viking Boat Museum, Oslo, Norway
Record Number: <12B1>
Description: Small fragment of silk embroidery decorated with two animals (Fig. 13). Although the ground fabric has disintegrated the stitch work has survived well, probably because it was so densely packed together when worked. Reasoning from the fineness of the embroidery and the evidence that vegetable fibres did not survive within the burial, the excavation report suggests the ground fabric was originally finely woven linen.

The design consists of two naturalistic animals, possibly lions or griffins, standing back-to-back with their heads turned so that they are looking at each other across their backs. The one on the left has a long unrolled tongue that extends behind and beyond its upturned tail. Falling down the back of its head is a mane-like swirl. The animal to the right is a similar motif but with its tongue positioned over the top of its tail and one front leg raised. Each animal is contained within its own roundel but adjacent to each other.

Between and below the two medallions is a line of embroidery that must have formed the start of another motif, which the report authors suggests is the tip of a leaf. Similarly a partial motif element can also be seen in the gap directly above the two roundels and to the left is the possible edge of another medallion.

Figure 13 shows the whole of the surviving piece is embroidered, including the background, with instances of stem stitch, satin stitch and possibly split stitch as well as couched work. Stitch measurements suggest the stitch work is not as fine as other Anglo-Saxon examples with measurements taken from the photograph suggesting stitch lengths of *c.* 5 mm long. Use of silk threads and the lustre they give to the finished design would make this a prized and valuable piece.

Running in a horizontal line above the roundels at the top edge of the fragment is a row of buttonhole stitch, which the report authors state was a red silk. Buttonhole stitch was used to finish edges of textiles decoratively throughout the early medieval

period. The decorative edging suggests this was part of a border or band of medallions inhabited by creatures.
Compare: Maaseik A, Maaseik B (infill and outline technique), Oseberg F

Oseberg F (Ose F)

Object Type: Fragment with embroidery
Date Late 8th–9th century
Find Date: 1904
Measurements: Width: 65 mm × height: 30 mm
Embroidery Stitches: Satin stitch, split stitch, stem stitch
Find Location: Ship burial at Oseberg farm, Norway
Present Location: Viking Boat Museum, Oslo, Norway
Record Number: <12B2>
Description: A fragment of silk embroidery that forms an ambiguous design, which the report authors surmise could be either a griffin inhabiting a trailing vine or a large flower growing from a trailing vine (Fig. 14). When viewing the photograph of the embroidery, a possible animal facing to the right with one of its forelegs raised is discernible. If this is a griffin inhabiting a vine-scroll, the fluidity of the design is similar in style to those embroidered on the Maaseik pieces (Pl. 19a, c). The report authors have pointed out that the shape of the leaves on the vine-scroll has counterparts in Anglo-Saxon manuscripts from the later 8th and early 9th centuries.

The embroidery technique in the photograph looks very similar to that of Oseberg E. The 'vine-scroll' is worked in stem stitch with larger areas filled with satin stitch and the background is completely covered with embroidery.
Compare: Oseberg E

Oseberg G (Ose G)

Object Type: Fragment with embroidery
Date: Late 8th–9th century
Find Date: 1904
Measurements: Width: 33 mm × height: 54 mm
Embroidery Stitches: Satin stitch(?), split stitch, stem stitch
Find Location: Ship burial at Oseberg farm, Norway
Present Location: Viking Boat Museum, Oslo, Norway
Record Number: <12B3>
Description: A section of silk embroidery stitched with a possible cross and stylised bird motifs (Fig. 15). Much of the surviving fragment is covered with the cross shape. It has been worked in a light silk thread and outlined in a darker one. At its centre is another cross-like motif that lies diagonally across beginning at each corner where the arms join what would be the main shaft. This part of the design, worked in the

darker silk thread, resembles a cord holding two pieces of wood together. Each of the four sectors of the design created by the arms of the cross appears to be inhabited by a stylised bird but apparent fading of the stitch work makes these difficult to decipher clearly (Fig. 35).

The embroidery appears to be worked in stem stitch and possibly split stitch with different shades of thread, demonstrating that a variety of coloured silks were originally used. The length of the stitches can be approximated from the photograph at 3–4 mm long. The standard of work appears to be high and consistent throughout the surviving fragment. The stylised cross motif with accompanying animal elements is seen throughout the early medieval period across Anglo-Saxon England art forms. Without further research, it is not possible to say whether this piece belongs with Oseberg E, F, H, I, and J.

Compare: N/A

Oseberg H (Ose H)

Object Type: Fragment with embroidery
Date: Late 8th–9th century
Find Date: 1904
Measurements: 45 mm × 40 mm
Embroidery Stitches: Satin stitch(?), split stitch, stem stitch
Find Location: Ship burial at Oseberg farm, Norway
Present Location: Viking Boat Museum, Oslo, Norway
Record Number: <12B4>
Description: A small fragment of embroidery covered with a geometric design worked in silk threads (Fig. 16). The report's authors state that the design is not clear but seems to consist of a border that surrounds a number of diagonal lines that form a lattice-like pattern, which accords with the rather indistinct photograph. The border has been subdivided into rectangles, two in each of the upright bars to the right, outlined with a darker thread.

Compare: Oseberg I

Oseberg I (Ose I)

Object Type: Fragment with embroidery
Date: Late 8th–9th century
Find Date: 1904
Measurements: 38 mm × 25 mm
Embroidery Stitches: Satin stitch(?), split stitch, stem stitch
Find Location: Ship burial at Oseberg farm, Norway
Present Location: Viking Boat Museum, Oslo, Norway
Record Number: <29B>

Description: Small fragment of silk embroidery worked with a geometric design (Fig. 17).

This piece is similar to Oseberg H in that its design is formed of geometric shapes stitched in lighter and darker threads, which are more easily distinguishable for this piece. In the photograph, the design is orientated so that the three bands lie vertically; but the embroidery could just as easily have been intended to lie horizontally. The two outer bands appear to be identical, with diamond-shaped motifs subdivided into bands outlined in the darker thread. The central band is divided from the outer ones by a narrow border, again outlined with the darker thread. Within the central section a vertical line of geometric motifs incorporates zig-zag lines. The design is symmetrical.

Stitches measured from the photograph are approximately 2 mm in length and this and the standard of work accords with Oseberg H. The report's authors suggest the two fragments belong together. While it is possible both pieces were made in the same place, the likeness may just be owing to the use of similar style and technique.

The design of these pieces suggests a border of some description but whether this was a self-contained piece, such as tablet-woven or detachable border, or whether it formed the border of a much larger textile is unclear.

Compare: Oseberg H

Oseberg J (Ose J)

Object Type: Fragment with embroidery
Date Late 8th–9th century
Find Date: 1904
Measurements: 12B5: height: 145 mm × width: 80 mm; 12B6: 60 mm × 55 mm; 140 mm × 100 mm; 90 mm × 115 mm
Embroidery Stitches: Satin stitch(?), split stitch(?), stem stitch(?)
Find Location: Ship burial at Oseberg farm, Norway
Present Location: Viking Boat Museum, Oslo, Norway
Record Number: <12B5>, <12L1>, <12L1b>, <12B6>, <12L2>
Description: A silk embroidery comprising five surviving fragments (Pl. 11, Figs 56, 57). The first piece, 12B5, is the largest. All the sections contain a spiral pattern with foliage motifs. Each spiral is made up of three bands that circle outwards until they split, continuing in three different directions and joining with bands from another spiral to create an interlocking pattern of circles. Each band is decorated with blocks of colour, light against dark, which are themselves made up of tiny arrowheads and diagonal lines. Each band is also outlined with what appear to be two rows of stitching.
Compare: N/A

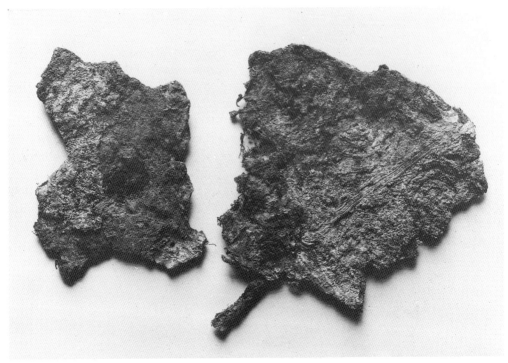

Figure 56. Oseberg J (large piece: 140 × 100 mm, small piece: 90 × 115 mm), © Museum of Cultural History, University of Oslo, Norway

Figure 57. Oseberg J: spiral (145 × 80 mm), © Museum of Cultural History, University of Oslo, Norway

Oseberg K (Ose K)

Object Type: Fragment with embroidery
Date: Late 8th–9th century
Find Date: 1904
Measurements: 45 mm × 10 mm
Embroidery Stitches: Stem stitch
Find Location: Ship burial at Oseberg farm, Norway
Present Location: Viking Boat Museum, Oslo, Norway
Record Number: <12G>
Description: The excavation report states only that this piece involves a piece of gathered fabric with the remains of embroidery worked in stem stitch and the remnants of possibly two other stitches. A skein of pale brown silk thread was associated with this item.
Compare: N/A

Oseberg L (Ose L)

Object Type: Fragment with embroidery
Date: Late 8th–9th century
Find Date: 1904
Measurements: 20 mm × 30 mm
Embroidery Stitches: Unidentified
Find Location: Ship burial at Oseberg farm, Norway
Present Location: Viking Boat Museum, Oslo, Norway
Record Number: <28B>
Description: The excavation report states that this piece comprises layers of a compressed bundle of embroidery. The design was stitched in a red and now grey silk in as yet unidentified stitches.
Compare: N/A

Sutton Hoo

Introduction

The embroideries found at Sutton Hoo were excavated from two different burial mounds. Sutton Hoo A was discovered during the 1939 excavations of the Ship Burial in Mound 1. Sutton Hoo B was found in a female burial in Mound 14 during the 1983–1991 excavations. Two fragmentary shoes that were once embroidered were also found in the Ship Burial but since they have no surviving embroidery they have been included in Table 3.
Sources for A: Crowfoot, 'The Textiles' (1983); Whiting, 'Dye Analysis' (1983)
Sources for B: Evans, 'Seventh-Century Assemblages' (2005); Walton Rogers, 'The Textiles from Mounds' (2005); Walton Rogers, *Cloth and Clothing* (2007)

Sutton Hoo A (Sut A)

Object Type: Seam fragment covered with embroidery
Date: 7th century
Find Date: 1939
Measurements: Length: 100 mm × left side width: 40 mm; right side width: 12 mm
Embroidery Stitches: Looped stitch
Find Location: Ship Burial, Mound 1, Sutton Hoo, Suffolk
Present Location: British Museum
Record Number: <SH9>, <1939,1010.184a>
Description: Wool embroidery covering a seam that joined two pieces of wool fabric together (Pl. 2a, Fig. 1). The embroidery was found in a bundle with other textiles that had been sandwiched between a mail-coat and silver vessels underneath an Anastasius dish. When Crowfoot took a section of the ground fabric apart, discerning that the textile's woven pattern was an irregular broken diamond, she also noted that the embroidery fibre was a single thread of wool similar to that used for the ground fabric. Results of tests on fibre samples at the Animal Breeding Research Organisation at the Roslin Institute, University of Edinburgh (1965–1970) showed the wool fibres were fine generalised medium wools with the occasional coarser fibre.

The embroidery and ground fabric are now uniformly brown and the original tones cannot be identified by eye. Samples were sent for dye analysis at the University of Bristol in 1970. This investigation found some samples contained widely varying quantities of indigotin, either from indigo or woad, which would have produced a blue colour. The variation was thought to be either due to the fact that only the warp threads were dyed, but some dye bled onto the weft threads, or because all the threads had been dyed but through varied decomposition the dye had faded at different rates on different threads. The analysis also found possible minute traces of madder, a red dye, on the weft threads of some samples. Crowfoot surmised that the seam was dyed blue but the report data is unclear on this point, and Crowfoot does not give a rationale.

The embroidery is a form of looped stitch, a stitch that was both decorative and functional (see discussion in Chapter 4). The version used on the Sutton Hoo fragment is one of the more basic examples; here the stitch reinforced the seam on to which it was sewn, making it stronger while keeping it flexible.

Although the Sutton Hoo piece is fragmentary and it is impossible to discern clearly its original function, Crowfoot suggested it could have been a cover for a pillow or bag, drawing on comparative evidence of embroidered seam on a pillow from Mammen in Denmark and another example at Welbeck Hill. More recent finds since Crowfoot's work, such as the Llangorse textile (1990) make it just as likely the Sutton Hoo textile was originally part of a garment. Since the pile in which it was found included textiles, was also associated with shoes and fragmented leather and was probably laid next to the deceased, we can hypothesise that the textiles were clothes of which the embroidered seam formed a part.
Compare: Dublin, Ingleby, Orkney, York B (looped stitch)

Sutton Hoo B (Sut B)

Object Type: Replaced textile with very fragmented embroidery
Date: Late 6th–early 7th century
Find Date: During 1983–1991 excavations
Measurements: Textile: 50 mm × 30 mm (embroidery covered most of the textile)
Embroidery Stitches: Running stitch, stem stitch
Find Location: Mound 14, Sutton Hoo, Suffolk
Present Location: British Museum
Record Number: <9a (50/4793)>, <1939,0411.2832>
Description: Embroidery and its associated textile replaced by the corroding iron of a now fragmented chatelaine, a set of metal rods that hung from the belt and from which objects were hung. The object was found in an inhumation burial, which had been robbed at some point in its history, and has been identified as that of a high-status female owing to the surviving artefacts, including the chatelaine, which is identified with the feminine gender, and found in graves dating from the 7th century. The piece is now in an extremely fragmentary and fragile state (Fig. 5).

The surviving metal replacement provides clear details of the material used and embroidery design and technique. According to Penelope Walton Rogers flax/hemp fibres were used to construct a linen ground fabric. Four sets of double parallel bars were embroidered on this (their length noted as 20 mm long), each set terminating in a roundel, with a further two fragmentary roundels discernible but not their associated bars.

Since the embroidered ground fabric was one of a number of replaced linen textiles and tablet-woven bands located in layers and folds around the chatelaine, Walton Rogers hypothesised as follows. The lowest layer was probably a long narrow sleeved chemise of medium-weight, linen, tabby woven fabric. Attached to the sleeve edge was a finer linen fabric on which the embroidery was worked, probably at the cuff. Above this layer was another fine linen fabric in the folds of which the metal chain and rods of the chatelaine were placed. On top of this lay a third layer of fine tabby woven fabric and cords that may have originated on tablet-woven bands. The layer in which the chatelaine was preserved was most likely sleeves of the garment and the tablet-woven bands were used to decorate the wide sleeve cuffs. Walton Rogers also suggested the embroidery may have been mimicking examples of cuffs constructed from tablet-woven bands and metal studs, fitting the pattern of development for dress accessories that took place during the 5th to the 6th centuries.
Compare: N/A

Utrecht (Utr)

Object Type: Relic pouch with embroidery
Date: 9th century
Find Date: no info
Measurements: Pouch height: 35 mm × width: 30 mm

Embroidery Stitches: Buttonhole stitch
Find Location: Utrecht Cathedral, Holland
Present Location: Lost/no record – last known location Utrecht Cathedral
Record Number: not supplied
Description: Relic pouch (or *bursa*), with embroidery (for an image see Battiscombe pl. XLIIIe). It was made from a gold-brocaded tablet-woven band stitched into a trapezoidal shape. Along the bottom of the pouch the tablet-woven band has been sewn with a functionally decorative buttonhole stitch, according to Granger-Taylor.

The pouch is associated with St Willibrord, an Anglo-Saxon missionary (658–739) who travelled to the Germanic kingdoms in the late 7th to early 8th centuries. It has been linked to Willibrord because it contains a bone fragment thought to be a relic from the saint himself. At some point the *bursa* and relic passed to Utrecht Cathedral, the saint's episcopal seat. It is probable that the relic came from Utrecht but the *bursa* is most likely Anglo-Saxon. St Willibrord was well known in England during his own lifetime and once he was canonised, pilgrimages to his tomb at Echternach would have been common. An Anglo-Saxon pilgrim may have carried a pouch housing important objects such as a bone from the saint. However, the relic and pouch may have been used to draw pilgrims to Utrecht. It may have been brought to Utrecht by a clergyman, or donated by a pilgrim as a gift in exchange for intercession on behalf of the giver. The *bursa* is small and was most likely recycled from a larger tablet-woven band. If this is so, the larger textile may have travelled across the Continent prior to becoming a pouch; and the embroidery may have been worked on the Continent.

Dominic Tweddle dated the band's production to the late 8th to early 9th century, so the functional embroidery must have been added to the band during or slightly after this, when it was converted from a band to relic pouch. The band is decorated with a vine scroll motif worked in two shades of red silk thread; the background was brocaded with a geometric design in gold thread. The buttonhole stitch was sewn in a contrasting lighter colour to those of the band, evidence of its decorative function. From the photograph that appears in the Battiscombe volume, stitches measure between 1 and 2.5 mm in length averaging approximately 2 mm. The stitches are regularly positioned 1 to 1.5 mm apart indicating the precision of the worker. According to Tweddle, the *bursa* was probably worn round the neck of the owner in a similar manner to the York A pouch. It is likely the relic was thought to bring protection to its wearer.
Compare: York A (object); Mitchell's Hill, Oseberg E, Sutton Hoo A, York B (functionally decorative embroidery)
Sources: Crowfoot, 'The Braids' (1956); Granger-Taylor, 'The Weft-Patterned Silks' (2002); Tweddle, 'Pouch as a Reliquary' (1989)

Winchester (Win)

Object Type: Part of leather vamp from item of footwear
Date: *c.* early 10th century

Find Date: During 1962–1971 excavations
Measurements: Shoe stitch length: 2 mm
Embroidery Stitches: Tunnel stitch(?)
Find Location: Castle Yard, Street 2, Winchester
Present Location: Hampshire Museum's Service Archive
Record Number: <CY OF 100>, catalogue number: <1852>
Description: Fragment of vamp (Fig. 9). According to Thornton, surviving evidence is not extensive enough to distinguish whether the vamp was from a boot or shoe. One edge of the leather has been torn while the other has been cut. A section of the lasting margin and two rows of stitching partially survive. The lasting margin consists of holes lying in a line, lengthways. At 45 degrees to the margin and the cut edge lie rows of decorative stitching, which probably formed part of a vamp stripe. Thornton observed that due to the angle of the surviving section, the stripe was most likely situated near the toe of the footwear. The two rows lie 3.5 mm apart with a stitch length of 2 mm. The stitches lie on the grain side of the leather and do not pierce the reverse side, suggesting the stitching is tunnel stitch.
Compare: Coventry, London B, London C (embroidered vamp)
Sources: Thornton, 'Shoes, Boots, and Shoe Repairs' (1990)

Worcester (Wor)

Object Type: Embroidered fragments
Date: 11th–12th century
Find Date: 1870
Measurements: Fragmentary silk strips: various lengths: 123–570 mm × various widths: 9–62 mm;
Wedges shapes: 1st: length: 179 mm × 81 mm wide at top and 91 mm wide at bottom (very fragmentary); 2nd: length: 173 mm × 64 mm at top (fragmentary at bottom); 3rd: length: 203 mm × 6 mm at top and 80 mm at bottom (my measurements differ slightly from those of Komatsu, probably because the pieces have been remounted)
Embroidery Stitches: Split stitch, stem stitch, underside couching
Find Location: Grave of a bishop, possibly William de Blois, in the Lady Chapel of Worcester Cathedral
Present Location: Worcester Cathedral Library
Record Number: none supplied
Description: Fragments of silk and silver-gilt embroidery (Pl. 1b, 28b, 29, 30a, 31a, c) which are believed to be remnants of a stole and maniple, as well as remnants of four wedge-shaped pieces and four fringes. These were discovered in 1870 while the paving in the Lady Chapel was being re-laid in the tomb (burial 18) ascribed to Bishop William de Blois (d. 1236), one of two (or possibly three, see below) lying under the paving in front of the chapel's altar. The embroideries were taken from the coffin before it was resealed and the effigies from tombs 18 and 20 were replaced in front of the altar. The embroideries were then moved and stored in the cathedral Chapter

House. In 1970 the effigy of burial 18 was moved to the side of the altar where it lies today. The archaeological report issued in 1974 noted that when the effigy was lifted there was no tomb beneath it. There is no record of what happened to the coffin and its remaining contents in 1870.

The earliest documentary evidence for the embroideries is in Green (1796). A piece of embroidery is mentioned in a footnote associated with a description of three burials in the Lady Chapel. The author had corresponded with a Dr Thomas who had been part of a group who opened what they thought was the tomb of King John and Dr Thomas had suggested investigating the tombs on either side: all three contained similar stone coffins buried to the same depth. At this date both burials 18 and 20 contained a skeleton with the remains of vestments, which Dr Thomas believed were bishops' vestments. Green's footnote says he had been given a small piece of embroidered cloth decorated with a figure accompanied by the word IEREMI (Jeremiah) and measuring five inches (114 mm) in length from the northern-most grave (burial 18). The fabric was brownish in colour with a fine texture, and parts of the design were worked in gold thread. This is not part of the collection held by the cathedral today but fits the style, design and colouration of surviving pieces.

According to St John Hope, the coffin and contents of this burial were not studied by experts at the time of their rediscovery in 1870, so this period yields no useful information: in particular the detail of whether the skeletons were clothed in the vestments or accompanied by them would change interpretations of their social and religious function. Hope noted that pieces of the vestments had been taken to the Chapter House and the larger segments of embroidery were preserved under glass, which suggests some of the embroidery may have been left in the coffin and that smaller fragments of embroidery may have been given away or not deemed worthy of preservation. At the time of the embroideries' removal to the Chapter House, Charles Henman was commissioned to draw the best preserved fragments but some of those segments are now housed with fragments he did not draw. Pieces are also known to have disappeared after being lent out to an exhibition in 1905 at the Burlington Fine Arts Club. Two fragments with embroidered crosses, one of which incorporated the head of an apostle and two letters from his name, 'MA', were not returned.

In 1938 Ada Grace Christie, who wrote the first technical analysis of the stitches and materials, concluded with Southwell and St John Hope that the segments had probably been part of a stole and maniple, most likely dated to the 11th to 12th centuries. A full scientific analysis was undertaken in 2006 to 2007 when the fragments were conserved at the Textile Conservation Centre in Southampton during which Miki Komatsu also conducted research on the pieces.

The probable stole consists of three strips made up of numerous fragments of varying sizes.

The pieces depict nine frames, each of which contains a haloed male figure holding a book (and in one case a sword). The figures range in height from 55 mm to 151 mm depending on how much has survived. The figures are identifiable as nine of the twelve apostles because six have their names at least partially extant, embroidered above their head: [IA]EOBBVS (St James), ANDRE (St Andrew), PAVLV (St Paul), TADEVS (St Thaddaeus), IHOAN (St John), and BART[HO]LOMEVS (St Bartholomew). Six face to the right and three to the left. Southwell suggested six of the apostles were placed on the left side of the stole facing to the right and vice versa, so the apostles faced each other across the body of the wearer. The Durham stole has figures facing alternately (Durham D), but Southwell's hypothesis, which placed St Paul prominently on the wearer's right and St Peter at a similar point on the left, is plausible.

Four of the six facing right survive practically whole; the bottom half of a second, the top half of a third and the torso of a fourth also remain. Of the surviving three figures facing left, one is practically whole, one formed of two pieces – the head and torso and the skirt of the clothing, and the third consists of the head and torso to just above the waistline. The figures fit a regular stylistic mould: an elongated, willowy body, a static pose with a slight bend at the knees on at least two and possibly more of the exemplars, and heads oval, with a pageboy haircut. Movement of the clothing is shown by a change of angle in the direction of the stitch work on the robe below the sash. The free hand not holding a book is either open palm up or closed with forefinger raised in blessing (St Paul is the exception, holding a sword, his traditional emblem). Feet are aligned in the same way. Eight of the nine are embroidered in three-quarter view with a ninth facing directly forward.

The probable maniple survives less well, in just three segments. The two largest pieces depict nearly complete figures. The third piece is small and shows only a tiny section of clothing around the sash. The figures are the same style and dimensions as the apostles on the stole. Two face right and one left. No inscriptions have survived. Christie suggested they represent the four major prophets as one holds an unfurled scroll, and a fragment that has since gone missing had the name Daniel embroidered on it. She also noted the piece given to Green included the name Jeremiah concluding with St John Hope before her, that the figures were Daniel, Ezekiel, Jeremiah and Isaiah.

Additionally, sections from four wedge-shaped pieces survive: one fairly complete, two partially complete and the fourth only a small fragment. Each shows a figure, a bishop and two kings, standing within an arched frame holding a symbol of office. The fourth fragment shows part of a skirt. The frames are each topped with a building surrounded by dots set in threes. Three pieces include lettering: NICOLAVS for the bishop, probably representing St Nicholas; ADELBERTVS above one king, probably for St Ethelbert who was king of East Anglia in the 8th century; and two letters, TO, which Christie took to represent St Thomas of Canterbury. The kings are rendered similarly to other Worcester figures, hold a sceptre and possible shield while the other hand reaches out in blessing. The bishop faces the viewer with his arms outstretched, a crosier on one hand (the other is now missing).

All the Worcester pieces are embroidered in the same style suggesting they formed a set, or were at least made in the same workshop setting. All were constructed from the same embroidery threads, use the same stitches and employ the same design features. The hair, clothing, halos, crowns, sword hilt, bishop's crosier, king's sceptre, lettering, buildings, dots and frames were worked in a silver-gilt metal thread which Komatsu's scientific analysis showed consisted of a metal strip constructed from two metals, a silver bottom layer gilded with a thin gold layer wound round a core of silk to create the thread. She also identified as linen the thread used to stitch the silver-gilt thread in place. Finer details – faces, necks, hands, feet, books, scrolls, sashes, and outlines on the clothing and figures – were all worked with non-metal threads identified by Komatsu as a silk thread in at least four colours: yellow, dark orange, light brown and dark brown. The design was worked on a red silk ground fabric: a tabby weave with 36 threads per cm on the warp and 34 picks per cm of the weft.

The surviving embroideries form a coherent set: the apostles on the stole and the prophets decorating the maniple. The four fringes were most likely attached to the ends of the stole and maniple: they are too narrow for the wedge-shaped pieces. The use of fringes to decorate the ends of stoles and maniples was common practice throughout the Middle Ages and can be seen in numerous manuscript illuminations. It is possible that the wedge-shaped pieces were tabs which intended to be placed back-to-back before adding them to the end of lappets hanging from the back of a bishop's mitre.

Compare: Durham F and G (silver-gilt and silk threads); Bayeux Tapestry, Durham C and D (style)

Sources: Christie, *English Medieval Embroidery* (1938); Gill, *Treatment Report* (unpublished); Green, *History and Antiquities of Worcester* (1796); Henman, 'Fragments of Ancient Vestments' (1871); Komatsu, 'Investigation of the Fragments of Liturgical Textiles' (2007); Perkins, 'Stole, Maniple and Four other Embroideries' (2007); Southwell, 'A Descriptive Account' (1913); St John Hope, *Proceedings* (1892); WCL A402(1), Cathedral Architect's Reports, Unpublished letter (1974)

Worthy Park (WPa)

Object Type: Replaced embroidery fragment
Date: Early/mid-6th–mid-7th century
Find Date: During 1962–1979 excavations
Measurements: Top layer of textile: 14 mm × 8 mm; Second (embroidered) layer: c. 25 mm × 15 mm; dimensions of embroidery not recorded
Embroidery Stitches: Satin stitch, stem stitch
Find Location: Inhumation Burial 75, Worthy Park, Kingsworthy, Hampshire
Present Location: Lost/no record
Record Number: 75(1)
Description: Replaced embroidery fragment found in contact with the blade of a knife (Fig. 7), located next to the left hip of the deceased where, as Walton Rogers points out, equipment was usually hung from a girdle or belt. The burial was undisturbed but badly degraded with very little surviving. Crowfoot understandably identified

the deceased as male on account of the knife, but the archaeological report is more hesitant, stating that the sex of the person is unknown.

Walton Rogers includes a diagram of the embroidery in her 2007 volume. Crowfoot examined the embroidery, which was found on the lower layer of fabric in a bundle of two pieces. She noted the ground fabric on which the design was worked was the finest piece of textile from the cemetery (20 × 20 threads per cm). Crowfoot suggested the embroidery may have been worked in silk due to the replaced thread having a soft appearance and variable spin direction, but acknowledged that no other silk had so far been found in a burial dating to this period. Crowfoot noted that even allowing for any shrinkage of the embroidery during the replacement process, the fineness of the embroidery, with stitches of satin and stem measuring just 1 mm, is remarkable.

She suggested that the embroidery's size means it was probably part of a small accessory, possibly a handkerchief or bag. It seems likely that it was housed on or in something hanging from the girdle or belt of the deceased. If housed in a container similar to the Kempston box, we would expect it to have survived. That the embroidery has undergone replacement from contact with the knife blade indicates it was hanging loose, possibly on the edge of a garment or a pouch hung at the waist. Due to the deterioration of some areas of embroidery and the fineness of the work, it was not possible to identify the design clearly. She tentatively suggested a leaf and scroll pattern.

Compare: N/A
Sources: Crowfoot, 'Chapter 5: The Textile Remains' (2003); Chadwick Hawkes and Grainger, *Anglo-Saxon Cemetery at Worthy Park* (2003); Walton Rogers, *Cloth and Clothing* (2007)

York

Introduction

York has been the focus of extensive systematic archaeological investigation since the middle of the 20th century. The topography and type of soils on which the city lies, have meant that large numbers of textile and leather fragments have survived, including embroidery produced with wool and silk thread.

In addition to the two surviving items included in the catalogue, another is in Table 2 and three possible embroideries in Table 3. The catalogue items were excavated at two different sites across York: 16–22 Coppergate and 28–29 High Ousegate. They are held either in the archive of York Archaeology Trust (YAT) or Yorkshire Museum but cannot currently be located.

Sources for A: Tweddle, 'Pouch as a Reliquary' (1989); Walton Rogers, *Textiles, Cordage and Raw Fibre* (1989)
Sources for B: Walton Rogers, 'Textiles, Cords, Animal Fibres' (2018)

York A (Yor A)

Object Type: Relic pouch
Date: c. 975
Find Date: During 1976–1982 excavations
Measurements: Pouch height: 33 mm × width: 25 mm–30 mm; Cross: 13 mm × 8 mm
Embroidery Stitches: Chain stitch
Find Location: Workshop/commercial setting: Tenement A, 16–22 Coppergate, York
Present Location: Unknown; last known location: Yorkshire Museum
Record Number: <1921>, [1977.7], <1408>
Description: Pouch constructed from two layers of (now pink) silk that form lining and outer sleeve embroidered with a cross (Fig. 3). The lining was made of two pieces of fawn-coloured silk in tabby weave stitched together forming a single rectangle measuring 50 mm × 22 mm. The outer silk was a six-sided fragment from a much larger cut-down piece of weft-faced compound twill silk. It is not patterned but dye analysis by Hofenk de Graaff at the Central Research Laboratory for Objects of Art and Science in Amsterdam, demonstrated the upper weft had been dyed with kermes and another, unknown, dye. As Tweddle noted, this means the silk was originally a rich purple colour, and he suggested that the symbolic value of expensive silk imported from a Byzantine or Islamic weaving centre, would have continued to adhere to the fabric after recycling, making it ideal for a relic pouch.

Walton Rogers states that only once the inner bag and its contents had been placed inside the outer pouch was it stitched together. This evidence, together with the embroidered Latin cross, suggest the contents of the bag constituted a relic. The cross was embroidered in the centre of the outer pouch in a now worn S-twist silk thread in chain stitch. It looks bulky on the small pouch, perhaps due to the thickness of the embroidery thread and the pouch's present condition. Yet average size of the eight surviving stitches on the archaeological drawing published in Walton Roger's report is 1.5 mm, so although Walton Rogers suggested the embroidery is crude workmanship compared to other silk embroidery of similar dates, stitch size is on a par with stitch work of other pieces in this catalogue, such as Kempston, which is considered fine work.

Tweddle believed the pouch's size meant it was intended to be worn round the neck of its owner to give protection. When the pouch was opened during conservation some probable vegetable fibres fell out, possibly either part of the relic or the wrapping that covered it. Tweddle suggests the fibres may have been part of a linen textile that could have come from clothing of a saint or else could have been a contact relic, that is, an item that had touched the body of a saint.
Compare: Utrecht (pouch)

York B (Yor B)

Object Type: Embroidered hem
Date: 10th century

Find Date: 2002
Measurements: Textile: 92 mm × 45 mm; Embroidery: approx. length: 160 mm × height: 5 mm
Embroidery Stitches: Looped stitch
Find Location: 28–29 High Ousegate, York
Present Location: Unknown; last known location: YAT archive
Record Number: SF00085(i)
Description: Cylindrical piece of fabric finished with embroidery at one end (Fig. 4).
 This object is a piece of repeat chevron twill fabric that has been sewn into a cylinder with a vertical seam. The shape of the piece suggests it is a sleeve with the embroidery forming the cuff that would have finished at the wrist. The cuff end is finished with a looped stitch used to bind the raw edge of the fabric whilst creating a decorative edge. The embroidery was worked in smoothly spun glossy wool constructed from a plied yarn made up of one Z-twisted and two S-twisted threads. The stitch itself is of the simple type used on Dublin and Sutton Hoo A. More complex forms of the stitch or different materials are used for Ingleby and Orkney.
Compare: Dublin, Ingleby, Orkney, Sutton Hoo A (looped stitch)

Appendix 2

Glossary of terminology

back stitch: (Fig. 58)

Figure 58. Back stitch, © Alexandra Lester-Makin

band: a narrow fabric that is produced by a non-weaving technique (*i.e.* off the loom), including plaits and other braids that do not contain a weft thread.
bast fibre: a fibre harvested from the inner stems of plants of the cellulosic fibre family, made into thread.
binding stitch: (Fig. 59)

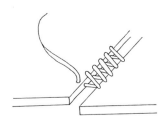

Figure 59. Binding stitch, © Alexandra Lester-Makin

brandea: a contact relic.
brocading: a supplementary weft thread used solely to create a woven pattern on a fabric or tablet-woven band.

broken diamond twill: (Fig. 60)

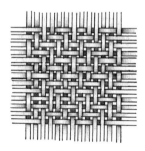

Figure 60. Broken diamond twill weave, © Alexandra Lester-Makin

buttonhole stitch: (Fig. 61)

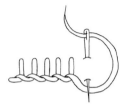

Figure 61. Buttonhole/blanket stitch, © Alexandra Lester-Makin

cast off: where the end of a thread that was used for sewing is finished off by working a number of small stitches to anchor it in place. The end of the thread will usually, but not always, be trimmed after it has been cast off.
cast on: where a thread is first attached to the ground fabric in order to start sewing or embroidering. The attachment usually takes the form of a knot or a couple of small stitches.
cellulosic fibres: fibres that come from plants and can be turned into thread.
chain stitch: (Fig. 62a, b)

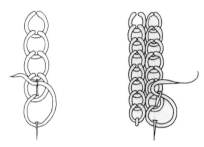

Figure 62. Chain stitch: a) single line; b) two rows, © Alexandra Lester-Makin

chatelaine: this is a set of short connector rods or metal linked chains that hang from a belt and from which items can be suspended. The hanging objects tend to be associated with the feminine gender and include keys and latch-lifters, amulets and copper alloy boxes, and textile-associated objects such as needle-cases and shears.
chevron twill: (Fig. 63)

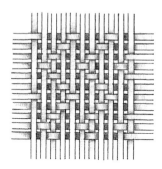

Figure 63. Chevron twill weave, © Alexandra Lester-Makin

couched work: (Fig. 64a, b)

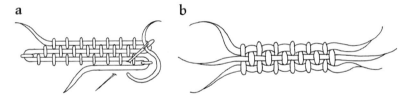

Figure 64. a) Couched metal threads; b) couched work, sometimes called tracey stitch, © Alexandra Lester-Makin

counted thread work: stitches are counted out as they are worked on an even-weave fabric or canvas.
diamond twill: (Fig. 65)

Figure 65. Diamond twill weave, © Alexandra Lester-Makin

embroidery: a decorative form of sewing. For the purposes of this book the definition is further delimited to cover only decorative work created with sewing thread or decoratively attaching other threads such as gold, to a pre-woven ground fabric.
fabric: a material constructed from fibres and/or threads that have been woven or matted together. Used predominantly for clothing, soft furnishings and decorative purposes.
fibre: a thread-like tissue taken from either an animal or plant that can be spun into a thread and/or woven into fabric or bands.
filling stitch: an embroidery or sewing stitch that is useful for covering flat areas of a design.
flax: the stem of this plant can be harvested for its bast fibres, called flax, and turned into linen fabric (also known as flax).
gold thread: in the early medieval period this thread was made by hammering out plates of gold into very thin sheets, then cutting them into narrow strips. Each strip was spun round a fibre core, normally a thread made of horse hair or silk, to produce the finished gold thread. It was used for embroidery or brocading and also woven into tablet-woven bands.
gold-work: embroidery worked with a gold thread.
ground fabric: the material on which embroidery is worked.
hem: the securing of an edge of fabric by over-sewing. The edge of the fabric may also be turned under so that raw edges are hidden from view and the folded edge sewn in place.
hemp: a bast fibre from the cannabis plant, which, when spun into thread, can be used for sewing or weaving.
herringbone stitch: (Fig. 66)

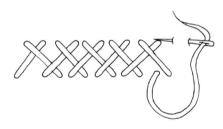

Figure 66. Herringbone stitch, © Alexandra Lester-Makin

herringbone twill: *see* **chevron twill**.
impression: an archaeological term to describe the image of one object that has been impressed upon another before the first object completed disintegrated thus leaving a negative image of its original form (*see also* **mineralised textile; replaced**).
incision: the place where the needle pierces the fabric during the sewing process.

laid-work: (Fig. 67)

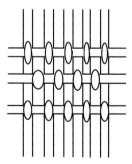

Figure 67. Laid-work, © Alexandra Lester-Makin

leather: the skin of an animal that is tanned and used to make products such as footwear and containers.
line stitch: a sewing or embroidery stitch worked in a line.
linen: a fabric woven from a spun flax or hemp fibre.
loom: a rectangular frame on which fabric is woven. The warp threads are attached to the frame and can be raised and lowered so that the fabric's pattern can be woven into it using the weft threads, which are passed under and over the warp threads as needed.
looped stitch: (Fig. 68a, b, c)

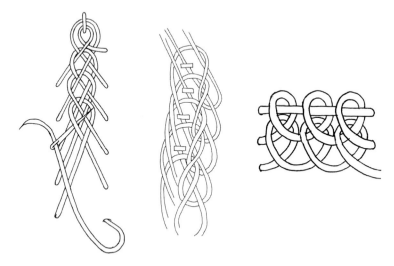

Figure 68. Looped stitch: a) Dublin, Sutton Hoo A, York; b) Orkney Hood; c) Ingleby, © Alexandra Lester-Makin

mineralised textiles (also known as **pseudomorphs**): an archaeological term to explain a change in material properties. As metals oxidise, their properties can seep into textiles and embroideries lying next to them. The properties then replace the original fibres as they disintegrate. The textile or embroidery survives as a mineralised object with none or very small amounts of the fibres remaining encased within the new formation (*see also* **replaced; impression**).

over-cast edge/whipped stitch: (Fig. 69)

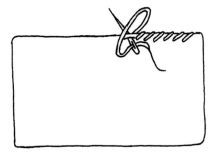

Figure 69. Overcast, © Alexandra Lester-Makin

over-sew stitch: (Fig. 70)

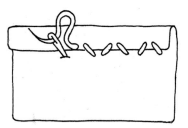

Figure 70. Over sew, © Alexandra Lester-Makin

plait(s): (Fig. 71)

Figure 71. Plait, © Alexandra Lester-Makin

Appendix 2

plait stitch: (Fig. 72a, b)

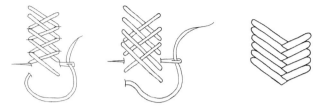

Figure 72. a) Plait stitch; b) raised plait stitch; c) a view of plait stitch and raised plait stitch when complete, © Alexandra Lester-Makin

ply/plied: (Fig. 73a, b, c)

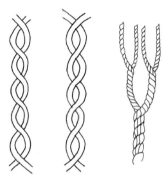

Figure 73. a) S-ply; b) Z-ply; c) plied thread, © Alexandra Lester-Makin

protein fibre: strands of keratin made of cells from the skin, wool or fur of an animal, forming a fibrous thread that can be harvested and used in textile production (*see also* **silk**).
pulled metal thread: a solid tube of metal that is pulled through smaller and smaller holes cut in a metal plate, until the thread is the diameter required for working.
radiocarbon dating: a method of determining the age of an object that contains organic material.
replaced: an archaeological term used to describe the survival of an object that is no longer made-up of its original materials, but has replaced or mineralised, normally by corroding metal (*see also* **mineralised textile; impression**).
running stitch: (Fig. 74)

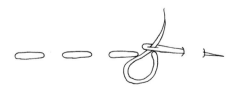

Figure 74. Running stitch, © Alexandra Lester-Makin

satin stitch: (Fig. 75)

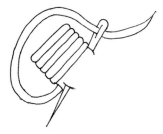

Figure 75. Satin stitch, © Alexandra Lester-Makin

seam: a join between two pieces of fabric that are sewn together.
seed stitch: (Fig. 76)

Figure 76. Seed stitch, © Alexandra Lester-Makin

selvedge: the side edges of a woven fabric created by the weft threads being looped from one row of weaving to the next.
silk: the thread and fabric made from the protein fibre of a cocoon made by a silkworm (*bombyx mori*).
silver-gilt thread: a metal thread similar to gold thread but made from silver that has been covered with a thin layer of gold. It took the place of gold thread in the later period, being cheaper to produce. The construction is the same as that for gold thread once the layer of gold has adhered to the silver (*see* **gold thread**).

slip hem stitch: (Fig. 77)

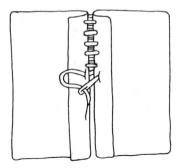

Figure 77. Slip hem, © Alexandra Lester-Makin

soumak: (Fig. 78a, b)

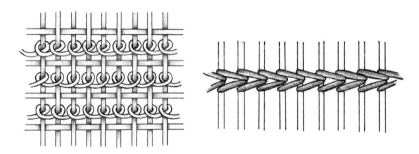

Figure 78. Soumak: a) thread woven through the ground fabric; b) compacted thread, © Alexandra Lester-Makin

spin/spun: (Fig. 79a, b)

Figure 79. a) S-spun; b) Z-spun, © Alexandra Lester-Makin

split stitch: (Fig. 80)

Figure 80. Split stitch, © Alexandra Lester-Makin

stem stitch: (Fig. 81a, b)

Figure 81. a) Stem stitch; b) stem stitch, two rows, © Alexandra Lester-Makin

stitch direction: the direction in which a stitch is sewn, for example from left-to-right or top-to-bottom.

tabby weave, sometimes called **plain weave**: (Fig. 82)

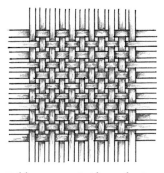

Figure 82. Tabby weave, © Alexandra Lester-Makin

tablet(s): a square or circular shaped piece of bone or wood that is thinned to the thickness of cardboard. Each tablet has two or four holes punched into it. To create a tablet woven band, the warp threads are passed through the holes repeatedly.

tablet-weaving/tablet-woven: a narrow band that has been woven 'off the loom' using tablets, which are twisted in various ways to form the fabric and its pattern.

thread: lengths of fibres twisted together by spinning and plying, or a continuous length of single fibre, unspun (such as silk). Both types of thread are used for functional and decorative sewing and weaving.

thread count: the number of threads counted per mm or cm in the weave of a fabric. If it is known which count is the warp and which the weft, the warp count is always noted first in British publications. Coarse textiles have lower thread counts while higher counts create finer fabrics or bands.

tunnel stitch: (Fig. 83a, b)

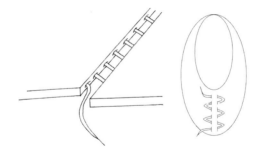

Figure 83. a) Tunnel stitch; b) tunnel stitch, front, © Alexandra Lester-Makin

twill weave: (Fig. 84a, b)

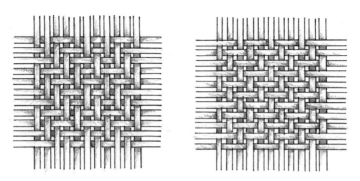

Figure 84. Twill weave a) 2×2; b) 2×1, © Alexandra Lester-Makin

twists: the motion used to move the tablets to produce the pattern on a tablet-woven band.
underside couching: (Fig. 85a, b, c)

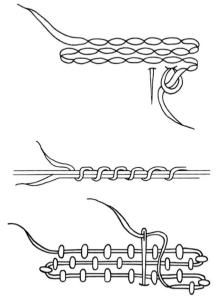

Figure 85. Underside couching: a) top view; b) side view; c) reverse view, © Alexandra Lester-Makin

unspun thread: one that has not been spun or plied.
warp: the fixed threads that are attached to the loom for weaving.
weave/woven: the motion of the warp and weft threads under and over each other in an ordered manner that follows a pattern and forms a textile (*see* **broken diamond twill, chevron twill, diamond twill, herringbone twill, tabby weave** and **twill weave** for the most popular types of weave during this period).
weft: the movable threads that create the pattern of a fabric by being woven under and over the warp threads.
wool: a fibre most commonly from the fleece of a sheep; also textiles and yarns made from the fleece of sheep.
worsted wool: a fabric made from a smooth yarn of long-stapled wool fleece that has been well-combed.
yarn: a continuous spun thread made from fibres that have been twisted and may also be plied.

Liturgical vestments

(Figs 86, 87, 88, 89, 90, 91)

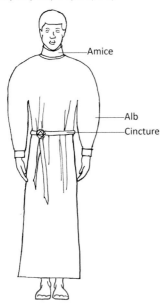

Figure 86. Amice and Alb, © *Alexandra Lester-Makin*

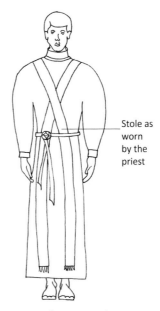

Figure 87. Stole: as worn by a priest, © *Alexandra Lester-Makin*

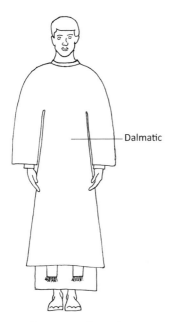

Figure 88. Dalmatic, © *Alexandra Lester-Makin*

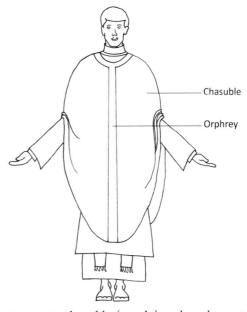

Figure 89. Chasuble (casula) and Orphrey, © *Alexandra Lester-Makin*

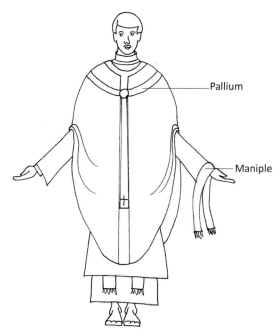

Figure 90. Pallum and Maniple, © Alexandra Lester-Makin

Figure 91. Stole: as worn by a deacon and bishop, © Alexandra Lester-Makin

Footwear terminology

(Figs 92, 93, 94a, 94b)

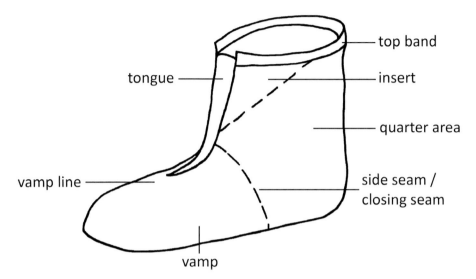

Figure 92. Boot, © Alexandra Lester-Makin

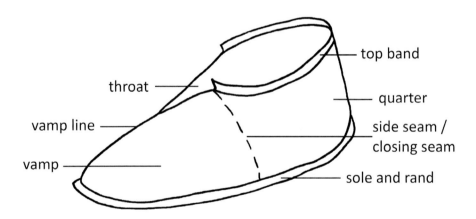

Figure 93. Shoe, © Alexandra Lester-Makin

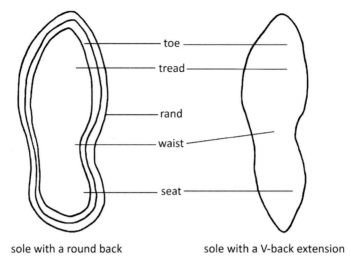

Figure 94. a) Normal sole; b) triangular sole, © *Alexandra Lester-Makin*

Terms in this Glossary have been variously compiled from: Butler, *The Batsford Encyclopaedia of Embroidery Stitches* (1982); Coatsworth, 'Embroideries from Tomb of Cuthbert' (2001); Coatsworth and Owen-Crocker, *Medieval Textiles of the British Isles* (2007); Cronyn, *Elements of Archaeological Conservation* (2001); Crowfoot, 'Appendix I. Objects found in Cremation Heath, Mound II' (1956); Crowfoot, 'The Textiles' (1983); Hald, *Ancient Danish Textiles from Bogs and Burials* (1980); Larratt Keefer, 'A Matter of Style: clerical vestments in the Anglo-Saxon church' (2007); Lester-Makin, 'Stitches' (2012); Lester-Makin, 'Stitches: filling stitches and other embroidery techniques' (2012); Lester-Makin, 'The Front Tells the Story' (2016); Morrell, *Guide to Structural Sewing* (1999); Mould, Carlisle and Cameron, 'Everyday Life' (2003); Plenderleith, 'The Stole and Maniples (a) the technique' (1956); Pritchard, 'Decoration' (2006); Textile Terms and Definitions website (2015).

Appendix 3

Table 1A-G: Table of surviving insular embroideries
Table 1A

Textile / catalogue code	Date	Bog	Cesspit / rubbish dump	Church treasury	Cloister	Coffin: clergy	Coffin: saint	Crannóg	Domestic setting	Inhumation
Alf	C5–6									•
Bay	C11			•						
Cov	L C11–E C12									
Dub	C10?		•							
Dur A	c. L C8?						•			
Dur B	c. L C8?						•			
Dur C	c. 910						•			
Dur D	c. 910						•			
Dur E	c. 910						•			
Dur F	1080–90?					•				
Dur G	1080–90?					•				
Ing	c. 878–900									
Kem	C7									
Lla	L C9–10							•		
Lon A	1050–1100									
Lon B	c. 1070s–1080s									
Lon C	c. 1070s–1080s									
Maa A	L C8–9			•						
Maa B	L C8–9			•						
Maa C	L C8–9			•						
Mil	C10			•						
Mit	C6?									
Ork	c. 250–615	•								
Ose A	L C8–9									
Ose B	L C8–9									
Ose C	L C8–9									
Ose D	L C8–9									
Ose E	L C8–9									
Ose F	L C8–9									
Ose G	L C8–9									
Ose H	L C8–9									
Ose I	L C8–9									
Ose J	L C8–9									
Ose K	L C8–9									
Ose L	L C8–9									
Sut A	E C7									
Sut B	E C7									
Utr	C9		•?							
Win	E C10								•	
Wor	C11–12				•					
WPa	M C6–M C7									•
Yor A	c. 975									
Yor B	C10									

Appendix 3

Textile / catalogue code	Discovery context							
	Inhumation: female	Mound	Mound: female	Open area within an urban setting	Pit	Ship burial: female	Ship burial: male	Workshop / commercial setting
Alf								
Bay								
Cov								•
Dub								
Dur A								
Dur B								
Dur C								
Dur D								
Dur E								
Dur F								
Dur G								
Ing		•						
Kem	•							
Lla								
Lon A					•			
Lon B				•				
Lon C				•				
Maa A								
Maa B								
Maa C								
Mil								
Mit	•							
Ork								
Ose A						•		
Ose B						•		
Ose C						•		
Ose D						•		
Ose E						•		
Ose F						•		
Ose G						•		
Ose H						•		
Ose I						•		
Ose J						•		
Ose K						•		
Ose L						•		
Sut A							•	
Sut B			•					
Utr								
Win								
Wor								
WPa								
Yor A								•
Yor B								•

Table 1B

Textile / catalogue code	Date	Bands	Embroidered edge	Embroidery with fabric fibre	Embroidered seam	Embroidered soumak-woven band	Embroidered tablet-woven band	Embroidered fabric	Fragment with no ground fabric	Girdle
Alf	C5–6							•		
Bay	C11									
Cov	L C11–E C12									
Dub	C10?				•					
Dur A	c. L C8?					•				
Dur B	c. L C8?						•			
Dur C	c. 910									•?
Dur D	c. 910									
Dur E	c. 910									
Dur F	1080–90?							•		
Dur G	1080–90?							•		
Ing	c. 878–900			•						
Kem	C7							•		
Lla	L C9–10							•		
Lon A	1050–1100									
Lon B	c. 1070s–1080s									
Lon C	c. 1070s–1080s									
Maa A	L C8–9	•								
Maa B	L C8–9	•								
Maa C	L C8–9	•								
Mil	C10			•						
Mit	C6?						•			
Ork	c. 250–615									
Ose A	L C8–9							•		
Ose B	L C8–9							•		
Ose C	L C8–9							•		
Ose D	L C8–9							•		
Ose E	L C8–9								•	
Ose F	L C8–9								•	
Ose G	L C8–9								•	
Ose H	L C8–9								•	
Ose I	L C8–9								•	
Ose J	L C8–9								•	
Ose K	L C8–9								•	
Ose L	L C8–9								•	
Sut A	E C7				•					
Sut B	E C7									
Utr	C9				•					
Win	E C10									
Wor	C11–12									
WPa	M C6–M C7							•		
Yor A	c. 975									
Yor B	C10			•						

Appendix 3

Textile / catalogue code	Object type										
	Hanging	Headdress	Hood	Maniple	Pouch (relic)	Shoe	Stole	Sleeve: cuff	Top-band	Tunic	Vamp: footwear
Alf											
Bay	•										
Cov						•					
Dub											
Dur A											
Dur B											
Dur C		•?									
Dur D							•				
Dur E				•							
Dur F											
Dur G											
Ing											
Kem											
Lla										•?	
Lon A									•		
Lon B											•
Lon C											•
Maa A											
Maa B											
Maa C											
Mil											
Mit											
Ork			•								
Ose A											
Ose B											
Ose C											
Ose D											
Ose E											
Ose F											
Ose G											
Ose H											
Ose I											
Ose J											
Ose K											
Ose L											
Sut A											
Sut B								•			
Utr											
Win											•
Wor							•				
WPa											
Yor A					•						
Yor B											

Table 1C
Condition and restoration

Textile / catalogue code	Date	Present condition					Restoration / conservation				
		Complete	Fragmentary	Mineralised	Partially complete	Replaced	Modern conservation	Modern stablisation	None	Partial conservation	Pre-1900 restoration
Alf	C5–6			•?		•?					
Bay	C11				•						
Cov	L C11–E C12				•					•?	
Dub	C10?		•						•		
Dur A	c. L C8?		•								•
Dur B	c. L C8?				•						•
Dur C	c. 910				•						•
Dur D	c. 910				•						•
Dur E	c. 910	•									•
Dur F	1080–90?		•								
Dur G	1080–90?		•								
Ing	c. 878–900		•						•		
Kem	C7		•					•			
Lla	L C9–10		•				•	•			
Lon A	1050–1100		•					•?	•?		
Lon B	c. 1070s–1080s		•					•?	•?		
Lon C	c. 1070s–1080s		•					•?	•?		
Maa A	L C8–9				•			•			
Maa B	L C8–9				•			•			
Maa C	L C8–9				•			•			
Mil	C10		•							•	•?
Mit	C6?		•								
Ork	c. 250–615				•			•			
Ose A	L C8–9		•								
Ose B	L C8–9		•								
Ose C	L C8–9		•								
Ose D	L C8–9		•								
Ose E	L C8–9		•								
Ose F	L C8–9		•								
Ose G	L C8–9		•								
Ose H	L C8–9		•								
Ose I	L C8–9		•								
Ose J	L C8–9		•								
Ose K	L C8–9		•								
Ose L	L C8–9		•								
Sut A	E C7		•						•		
Sut B	E C7					•			•		
Utr	C9	•							•		
Win	E C10		•						•		
Wor	C11–12				•			•			
WPa	M C6–M C7					•					
Yor A	c. 975				•				•		
Yor B	C10		•						•		

Appendix 3

Table 1D

Textile / catalogue code	Date	Ground fabric					Weave						
		Leather	Linen	None	Silk	Wool	Broken chevron twill	Broken diamond twill	Compound twill	Chevron twill	Herringbone twill	Tabby weave	Tablet-woven
Alf	C5–6		•?										
Bay	C11		•									•	
Cov	L C11–E C12	•											
Dub	C10?				•								
Dur A	c. L C8?				•							•	
Dur B	c. L C8?				•							•	•
Dur C	c. 910				•							•	
Dur D	c. 910				•							•	
Dur E	c. 910				•							•	
Dur F	1080–90?				•								
Dur G	1080–90?				•								
Ing	c. 878–900				•?								
Kem	C7					•		•					
Lla	L C9–10		•									•	
Lon A	1050–1100	•											
Lon B	c. 1070s–1080s	•											
Lon C	c. 1070s–1080s	•											
Maa A	L C8–9		•									•	
Maa B	L C8–9		•									•	
Maa C	L C8–9		•									•	
Mil	C10				•								
Mit	C6?					•							•
Ork	c. 250–615					•	•						•
Ose A	L C8–9					•						•	
Ose B	L C8–9					•						•	
Ose C	L C8–9					•						•	
Ose D	L C8–9					•						•	
Ose E	L C8–9				•								
Ose F	L C8–9				•								
Ose G	L C8–9				•								
Ose H	L C8–9				•								
Ose I	L C8–9				•								
Ose J	L C8–9				•								
Ose K	L C8–9				•								
Ose L	L C8–9				•								
Sut A	E C7					•		•					
Sut B	E C7		•									•	
Utr	C9				•?								•
Win	E C10	•											
Wor	C11–12				•							•	
WPa	M C6–M C7				•							•	
Yor A	c. 975				•				•				
Yor B	C10					•					• warp		

Table 1D (Continued)

Textile / catalogue code	Gauge				Ground fabric colour													
	Not known	Warp	Weft	Warp and weft not stipulated	Blue	Brown	Brown-purple	Buff	Fawn	No dye	No dye detected	Pink	Purple	Red	Red-brown	Violet	White	Yellow
Alf																		
Bay								•										
Cov								•										
Dub																		
Dur A					•	light	•						brown	•	•			gold
Dur B														•				•
Dur C		56	56														•	
Dur D		56	56														•	
Dur E		64?	64?														•	
Dur F																		
Dur G																		
Ing																		
Kem				28 / 22									•					
Lla		25	25															
Lon A										•								
Lon B										•								
Lon C										•								
Maa A				24 × 26						•								
Maa B				26 × 20						•								
Maa C				24 × 20	P													
Mil		? 15-6	? 10					warp								weft		
Mit																		•?
Ork		7-10	17-19							•								
Ose A														•				
Ose B														•				
Ose C														•				
Ose D														•				
Ose E	•																	
Ose F	•																	
Ose G	•																	
Ose H	•																	
Ose I	•																	
Ose J	•																	
Ose K	•																	
Ose L	•																	
Sut A		17-19	13-15		•?**									•?*				
Sut B				30 × 28								•						
Utr	•													•				
Win																		
Wor		36	34											•				
WPa		22	22															
Yor A	•								•				•					
Yor B				Z12 / S10							•							

P = painted; * weft?; ** weft-D? warp-L?

Appendix 3

Table 1E

Textile / catalogue code	Date	Embroidery thread							Spin / ply	Thickness of		
		Gold	*Linen*	*Replaced*	*Silk*	*Silver*	*Silver gilt*	*Wool*	*S / Z*	*Fine*	*Medium*	*Very fine*
Alf	C5–6			•								
Bay	C11		•					•				
Cov	L C11–E C12							•				
Dub	C10?							•				
Dur A	*c.* L C8?				•				S twist	•		
Dur B	*c.* L C8?				•				S			•
Dur C	*c.* 910	•			•				S / Z	•		•
Dur D	*c.* 910	•			•				S / Z	•		•
Dur E	*c.* 910	•			•				S / Z	•		•
Dur F	1080–90?				•		•					
Dur G	1080–90?				•		•					
Ing	*c.* 878–900					•						
Kem	C7							•	Z / S			•
Lla	L C9–10				•				2P & 1P			•
Lon A	1050–1100							•				
Lon B	*c.* 1070s–1080s				•?							
Lon C	*c.* 1070s–1080s											
Maa A	L C8–9	•			•				Z, S, untwisted	•		
Maa B	L C8–9	•			•				Z, S, untwisted	•		
Maa C	L C8–9	•			•				Z, S, untwisted	•		
Mil	C10	•			•				S plied	•	•	
Mit	C6?							•	2P		•	
Ork	*c.* 250–615							•	2P, lightly spun	•		
Ose A	L C8–9							•		•		
Ose B	L C8–9											
Ose C	L C8–9							•	2P	•		
Ose D	L C8–9							•	Z spun	•		
Ose E	L C8–9				•							•
Ose F	L C8–9				•							•
Ose G	L C8–9				•							•
Ose H	L C8–9				•							•
Ose I	L C8–9				•							•
Ose J	L C8–9				•							•
Ose K	L C8–9				•							•
Ose L	L C8–9				•							•
Sut A	E C7							•		•		
Sut B	E C7							•	plied - Z2S			•
Utr	C9				•?					•		
Win	E C10									•		
Wor	C11–12				•		•		2S	•		•
Wpa	M C6–M C7				•?				variable spun, S?			•
Yor A	*c.* 975				•				none	0.5 mm		
Yor B	C10							•	Z2S	1.5		

Table 1E (Continued)

Textile / catalogue code	Embroidery thread — Embroidery thread colour																				
	Beige	Blue	Brown	Brown-purple	Gold	Green	Grey	Fawn	No dye	No dye detected	Orange	Pink	Purple	Red	Red-brown	Red-purple	Silver	Stained	Violet	White	Yellow
Alf																					
Bay																					
Cov																					
Dub																					
Dur A		•	light	•								•	•	•							•
Dur B		•?	light?	•?								•?	•	•?							•?
Dur C					•								•								
Dur D			dark		•	dark, •	•						•		•						
Dur E			dark		•	dark, •	•						•		•						
Dur F																					
Dur G																					
Ing																			•		
Kem		•?				•?								•					•?		•?
Lla																					
Lon A													•								
Lon B																					
Lon C																					
Maa A	•	L & D			•	•								•							•
Maa B	•	L & D			•	•								•							•
Maa C	•	L & D			•	•								•							•
Mil		D?			•									•					?		
Mit								•?													•?
Ork			dark																		
Ose A									•												
Ose B																					
Ose C									•												
Ose D														•						•	
Ose E																		•			
Ose F																		•			
Ose G																		•			
Ose H																		•			
Ose I																		•			
Ose J					•?								•		•			•			•?
Ose K			pale											•							
Ose L							•						•					•			
Sut A		•																			
Sut B																					
Utr												•?									
Win																					
Wor		• & D									dark										•
Wpa																					
Yor A												•?									
Yor B							•														

Table 1F

Textile / catalogue code	Date	Buttonhole stitch	Chain stitch	Couched	Counted	Laid-work	Looped stitch	Raised plait	Running stitch	Satin stitch	Split stitch	Stem stitch	Tunnel stitch	Underside couching	Unidentified
Alf	C5–6														
Bay	C11		•			•					•	•			
Cov	L C11–E C12							•							
Dub	C10?						•								
Dur A	c. L C8?									•?					
Dur B	c. L C8?	•?								•?					
Dur C	c. 910			•								•			
Dur D	c. 910			•							•	•			
Dur E	c. 910			•							•	•			
Dur F	1080–90?			•		•				•		•			
Dur G	1080–90?											•		•	
Ing	c. 878–900						•								
Kem	C7		•?								•	•			
Lla	L C9–10				•							•			
Lon A	1050–1100								•?				•?		
Lon B	c. 1070s–1080s												•		
Lon C	c. 1070s–1080s												•?		
Maa A	L C8–9			•								•			
Maa B	L C8–9			•								•			
Maa C	L C8–9			•								•			
Mil	C10			•								•			
Mit	C6?	•													
Ork	c. 250–615		•			•									
Ose A	L C8–9			•?								•?			
Ose B	L C8–9											•?			
Ose C	L C8–9														
Ose D	L C8–9											•?			
Ose E	L C8–9	•		•						•	•	•			
Ose F	L C8–9									•	•	•			
Ose G	L C8–9									•?	•	•			
Ose H	L C8–9									•?	•	•			
Ose I	L C8–9									•?	•	•?			
Ose J	L C8–9									•?	•?	•?			
Ose K	L C8–9											•			
Ose L	L C8–9														•
Sut A	E C7					•									
Sut B	E C7								•			•			
Utr	C9	•													
Win	E C10												•?		
Wor	C11–12										•	•		•	
WPa	M C6–M C7								•			•			
Yor A	c. 975		•												
Yor B	C10						•								

Table 1G

Textile / catalogue code	Date	Animals	Arcades	Bars	Buildings	Chevrons	Cross	Dots	Figures	Flowers	Foliage	Fruit
Alf	C5–6											
Bay	C11	•		•	•	•			•	•	•	
Cov	L C11–E C12											
Dub	C10?											
Dur A	c. L C8?			•								
Dur B	c. L C8?			•								
Dur C	c. 910	•								•	•	•
Dur D	c. 910								•		•	
Dur E	c. 910								•		•	
Dur F	1080–90?							•			•	
Dur G	1080–90?	•									•	
Ing	c. 878–900											
Kem	C7	•										
Lla	L C9–10	•									•	
Lon A	1050–1100											
Lon B	c. 1070s–1080s											
Lon C	c. 1070s–1080s											
Maa A	L C8–9	•	•			•					•	
Maa B	L C8–9	•									•	
Maa C	L C8–9										•	
Mil	C10	•						•			•	
Mit	C6?											
Ork	c. 250–615											
Ose A	L C8–9	•?										
Ose B	L C8–9											
Ose C	L C8–9											
Ose D	L C8–9											
Ose E	L C8–9	•									•	
Ose F	L C8–9	•?							•		•	
Ose G	L C8–9	•					•					
Ose H	L C8–9			•?								
Ose I	L C8–9			•								
Ose J	L C8–9			•		•					•?	
Ose K	L C8–9											
Ose L	L C8–9											
Sut A	E C7											
Sut B	E C7			•								
Utr	C9											
Win	E C10											
Wor	C11–12			•?	•			•	•			
WPa	M C6–M C7										•?	
Yor A	c. 975						•					
Yor B	C10											

Appendix 3

Textile / catalogue code	Geometric	Inscription	Interlace	Knot work	None detected	Religious symbols	Rings	Roundels	Scroll border	Spandrels	Spiral	Stem	Stripe (vamp)	Symbol of office / status	Vine	Vine scroll
Alf	•															
Bay		•				•										
Cov													•			
Dub				•												
Dur A																
Dur B																
Dur C												•				
Dur D		•				•								•		
Dur E		•				•								•		
Dur F																
Dur G							•									
Ing					•											
Kem				•												
Lla	•								•							
Lon A					•											
Lon B													•			
Lon C													•			
Maa A	•		•							•						
Maa B								•								
Maa C	•															•
Mil	•?															•
Mit					•											
Ork					•											
Ose A	•?															
Ose B																
Ose C							•									
Ose D					•											
Ose E								•								
Ose F															•	
Ose G																
Ose H	•															
Ose I	•															
Ose J											•				•?	
Ose K					•											
Ose L					•											
Sut A					•											
Sut B								•								
Utr					•											
Win													•?			
Wor		•												•		
WPa									•?							
Yor A																
Yor B					•											

Design elements

Table 2: Loose gold thread

Catalogue code	Date	Discovery context						Present condition			Thread constituents		
		Cesspit / rubbish dump	Domestic / workshop setting	Open area within an urban setting	Prestigious grave: female	Prestigious grave: male	Road	Core and metal foil	Metal foil only	Poor	Silk	Animal hair	Gold
London Ga	c. 1060–early 1100s						•			•			•
London Gb	c. 1070s–1080s			•						•			•
Repton Ga									•				•
Repton Gb					•				•				•
Repton Gc						•			•				•
Southampton Ga	c. 8th–9th centuries	•											
York Ga	c. 975–middle–late 11th century		•										•
Winchester Ga	10th century					•			•				•

Catalogue numbering is author's own and indicates find location. As discussed in the introduction, these items are not included in the catalogue or embroideries, but they give the reader a useful indication of other finds potentially associated with embroidery.

The data used to compile Table 2: Loose Gold Thread has been taken from: Biddle, pers. comm. (2014); Biddle and Kjølbye-Biddle, 'Repton and the Great Heathen Army, 873-4' (2001); Crowfoot, 'Personal Possessions: textiles' (1990); Egan, '8.8 Accessioned Finds' (2007); Pritchard, pers. comm. (2015); Walton, *Textiles, Cordage and Raw Fibre* (1989); Walton Rogers, *Textile Production* (1997); Walton Rogers, 'Gold Thread' (2005).

Appendix 3

Table 3: Pieces with no surviving thread

Catalogue code	Date	Publication code	Building	Cesspit / rubbish dump	Domestic setting	Inhumation: male	Open area within an urban setting	Road	Ship burial: male	Structure
Buc	C5–6	spearhead 4/1				•				
Lon Na	900–970	L62					•			
Lon Nb	C10?	SWA81						•?		
Lon Nc	L. C10	L31				•				
Lon Nd	L. C10	L35				•				
Lon Ne	L. C10	L75				•				
Lon Nf	L. C10	L79				•				
Lon Ng	c. 1000–1050	L12						•		
Lon Nh	c. 1000–1050	L13						•		
Lon Ni	c. 1000–1050	L14						•		
Lon Nj	c. 1000–1050	L16						•		
Lon Nk	c. 1000–1050	L38						•		
Lon Nl	c. 1000–1050	L40						•		
Lon Nm	c. 1000–1050	L85	•							
Lon Nn	c. 1000–1050	L110						•		
Lon No	c. 1000–1050	L111						•		
Lon Np	c. 1000–1050	L131						•		
Lon Nq	c. 1000–1050	L134						•		
Lon Nr	c. 1000–1050	L139						•		
Lon Ns	c. 1050–1070	L55					•			
Lon Nt	c. 1050–1070	L64					•			
Lon Nu	c. 1050–1100	L24					•			
Lon Nv	c. 1050–1100	L158					•			
Lon Nw	c. 1060–1100	L53								•
Lon Nx	c. 1060–1100	L84								•
Lon Ny	c. 1060–1100	L85								•
Lon Nz	c. 1060–1100	L86								•
Lon Naa	c. 1070s–1080s	L51					•			
Lon Nab	c. 1070s–1080s	L68					•			
Lon Nac	L C11	L83						•		
Lon Nad	?	NFW74								
Sut Na	E C7	173.5a							•	
Sut Nb	E C7	173.9a & b							•	
Yor Na	M–L C9–E C10	15453			•					
Yor Nb	M–L C9–E C10	15524			•					
Yor Nc	L C9–E C10	651			•					
Yor Nd	L C11–M C12	15459			•					

Table 3: Pieces with no surviving thread (Continued)

Catalogue code	Object							Present condition				
	Fabric with holes	Leather with holes	Replaced	Shoe/boot	Shoe upper	Top band	Vamp	Complete	Fragmentary	Impression	Incomplete	Mineralised	Near complete
Buc			•									•	
Lon Na				•			•				•		
Lon Nb					•		•				•		
Lon Nc				•							•		
Lon Nd				•				•					
Lon Ne				•				•					
Lon Nf				•							•		
Lon Ng				•									•
Lon Nh							•				•		
Lon Ni							•				•		
Lon Nj							•				•		
Lon Nk				•							•		
Lon Nl				•							•		
Lon Nm				•							•		
Lon Nn					•						•		
Lon No				•							•		
Lon Np				•							•		
Lon Nq					•						•		
Lon Nr				x2							•		
Lon Ns				•							•		
Lon Nt					•						•		
Lon Nu				•							•		
Lon Nv				•				•					
Lon Nw							•				•		
Lon Nx					•						•		
Lon Ny							•				•		
Lon Nz							•				•		
Lon Naa					•						•		
Lon Nab							•				•		
Lon Nac							•				•		
Lon Nad						•					•?		
Sut Na		•		•					very				
Sut Nb		•					•		•				
Yor Na		•							•				
Yor Nb							•		•				
Yor Nc	•												
Yor Nd					•				•				

Appendix 3

Catalogue code	Restoration / conservation			Ground fabric		Thread hole thickness			
	Modern conservation	*None*	*Partial conservation*	*Leather*	*None*	*Fine*	*Medium*	*Not known*	*Thick*
Buc		•			tabby				•
Lon Na		•		•				•	
Lon Nb		•		•				•	
Lon Nc		•		•				•	
Lon Nd		•		•				•	
Lon Ne		•		•				•	
Lon Nf		•		•				•	
Lon Ng		•		•				•	
Lon Nh		•		•				•	
Lon Ni		•		•				•	
Lon Nj		•		•				•	
Lon Nk		•		•				•	
Lon Nl		•		•				•	
Lon Nm		•		•				•	
Lon Nn		•		•				•	
Lon No		•		•				•	
Lon Np		•		•				•	
Lon Nq		•		•				•	
Lon Nr		•		•				•	
Lon Ns		•		•				•	
Lon Nt		•		•				•	
Lon Nu		•		•				•	
Lon Nv		•		•				•	
Lon Nw		•		•		•			
Lon Nx		•		•		•			
Lon Ny		•		•				•	
Lon Nz		•		•				•	
Lon Naa		•		•				•	
Lon Nab		•		•				•	
Lon Nac		•		•				•	
Lon Nad		•		•				•	
Sut Na			•			•			
Sut Nb		•				•			
Yor Na		•?		•					
Yor Nb		•?		•					
Yor Nc		•?							
Yor Nd		•?		•					

Catalogue code	Stitch							Design element			Vamp strip(s)
	Herringbone	Not known	Running	Satin	Soumak	Tunnel stitch	Decorative reinforcement	Geometric	None detected	Not known	
Buc	•?				•?				•		
Lon Na											•
Lon Nb	•										•
Lon Nc	•										•
Lon Nd	•										•
Lon Ne	•										•
Lon Nf	•										•
Lon Ng	•										•
Lon Nh	•										•
Lon Ni	•										•
Lon Nj	•										•
Lon Nk	•										•
Lon Nl	•										•
Lon Nm	•										•
Lon Nn	•										•
Lon No	•						•				
Lon Np	•										•
Lon Nq	•										•
Lon Nr	•										x1
Lon Ns						•?					•
Lon Nt						•?					•
Lon Nu						•?					•
Lon Nv		•									•
Lon Nw		•									•
Lon Nx						•?					•
Lon Ny		•									•
Lon Nz						•?					•
Lon Naa						•?					•
Lon Nab	•										•
Lon Nac						•?					•
Lon Nad	•								•		
Sut Na				•				•			
Sut Nb			•						•		
Yor Na											•
Yor Nb						•					•
Yor Nc									•?		
Yor Nd						•					•

Catalogue numbering is author's own and indicates find location. As discussed in the introduction, these items are not included in the catalogue or embroideries, but they give readers a useful indication of other finds potentially associated with embroidery.

The data used to compile Table 3: Pieces with no Surviving Thread has been taken from: Crowfoot, 'The Textiles' (1987); East, 'The Shoes' (1983); MacConnoran with Nailer, 'Complete Catalogue of Leather Items' (CD: 2007); MacConnoran with Nailer, '8.15 Leather Items' (2007); MacConnoran, 'Complete Catalogue of Selected Medieval Fragments from 1 Poultry' (CD: 2011); Mould, Carlisle and Cameron, 'Catalogue' (2003).

Appendix 4

Chronological chart

Historical Event	Year by century (AD)	Embroidery
Uí Liatháin and *Laigin* migrated from Ireland to Wales.	mid-3rd–mid-6th	Orkney Hood
Dál Riata migrate from Ireland to Scotland.		Alfristan
Christianity arrives in Ireland and Wales.		
Angles, Saxons, Jutes arrive in England.		
Supremacy of the Kingdom of Gwynedd in north and east Wales.	6th	Mitchell's Hill
St Augustine arrives in Kent.		
Celtic Christian monastery founded on Iona.		
Anglo-Saxon conversion to Christianity begins.	6th–7th	Worthy Park
Northumbrian supremacy.	7th	Sutton Hoo A
Synod of Whitby leads to Anglo-Saxon Kingdoms following the Roman Church.		Sutton Hoo B
		Kempston
Mercian supremacy is solidified.	8th	Durham A
Offa's dyke is constructed.		Durham B
Viking raids across England, Ireland, Scotland and Wales.	8th–9th	Maaseik A
		Maaseik B
		Maaseik C
		Oseberg A
		Oseberg B

(Continued)

Chronological chart (Continued)

Historical Event	Year by century (AD)	Embroidery
		Oseberg C
		Oseberg D
		Oseberg E
		Oseberg F
		Oseberg G
		Oseberg H
		Oseberg I
		Oseberg J
		Oseberg K
		Oseberg L
Mercian supremacy supplanted by Kingdom of Wessex.	9th	Ingleby
Vikings settle in England; 'Danelaw' established.		Utrecht
Monastic towns develop in Ireland.		
Viking Dublin established.		
Anglo-Saxons continue to resist Vikings.	9th–10th	Llangorse
Anglo-Saxons defeat Vikings.	10th	Dublin(?)
Unified England is established.		Durham C
Irish reduce Viking power then conquer Dublin.		Durham D
Dyfed-Deheubarth laws established in Wales.		Durham E
Benedictine Religious Reforms.		Milan
		Winchester
		York A
		York B
Battle of Hastings: the Normans conquer England	11th	Bayeux Tapestry
		Durham F
		Durham G
		London B
		London C
Normans cement power in England	11th–12th	Coventry
		London A
		Worcester

Bibliography

Primary sources

Arnold, T. (ed.) (1882) 'Historia de sancto Cuthberto', *Symeonis monachi Opera Omnia*, Rolls Series 1. London: Longman.

Barlow, F. (ed. and trans.) (1962) *The Life of King Edward, who rests at Westminster*, Second edition. London: Nelson and Sons.

Binchy, D.A. (ed.) (1978) *Corpus Iuris Hibernici: ad fidem codicum manuscriptorum recognovit*, vol. 2. Dublin: Dublin Institute for Advanced Studies.

Blake, E.O. (ed.) (1962) *Liber Eliensis*. London: Royal Historical Society.

Campbell, A. (ed.) (1950) *Frithegodi monachi Breuiloquium vitae Beati Wilfredi = et Wulfstani cantoris Narratio metrica de Sancto Swithuno*. Turici: In Aedibus Thesauri Mundi.

Colgrave, B. and R.A. Mynors (eds) (1969) *Bede's Ecclesiastical History of the English People*. Oxford: Clarendon.

Davis, R. (trans.) (1995) *The Lives of the Ninth-Century Popes (Liber Pontificalis)*. Liverpool: Liverpool University Press.

de Gray Birch, W. (ed.) (1893) *Cartularium Saxonicum: a collection of charters relating to Anglo-Saxon history*, vol. 1. London: Chas. J. Clark.

Dümmler, E. (ed.) (1892) 'S. Bonifatii et Lulli epistolae', *Monumenta Germaniae Historica, Epistolae 3, Merovingici et Karolini Aevi, I*. Berlin: Weidmann, 215–433.

Fairweather, J. (trans.) (2005) *Liber Eliensis: a history of the Isle of Ely from the seventh century to the twelfth*. Woodbridge: Boydell Press.

Giles, [Rev.] J.A. (1844) *The Complete Works of Venerable Bede, In the Original Latin, Accompanied by a new English Translation: vol. IX Commentaries on the Scriptures*. London: Whittaker & co.

Goscelin (1938) 'La Légende de S. Édith', *Analecta Bollandiana*, vol. 56. Bruxelles: Société des Bollandistes.

Gwara, S. and R. Ehwald (eds) (2001) *Corpus Christianorum Series Latina: Aldhelmi Malmesbiriensis prosa de Virginitate: cum glosa latina atque anglosaxonica*, cxxiv A, chapter LV.

Keynes, S. (ed.) (1996) *The Liber Vitae of the New Minster and Hyde Abbey Winchester, British Library Stowe 944*. Copenhagen: Rosenkilde and Bagger.

Musset, L. (1967) *Les Actes de Guillaume le Conquérant et de la Reine Mathilde pour les Abbayes Caennaises*. Caen: Société des Antiquaires de Normandie.
Mynors, R.A.B., R.M. Thomson and M. Winterbottom (eds) (1998) *William of Malmesbury, Gesta Regum Anglorum: The History of the English Kings*. Oxford: Clarendon Press.
Riley, H.T. (trans.) (1854) *Ingulph's chronicle of the abbey of Croyland: with the continuations by Peter of Blois and anonymous writers*. London: H.G. Bohn.
Robertson, A.J. (1956) *Anglo-Saxon Charters*, Second edition. Cambridge: Cambridge University Press (repr. 2009).
Rollason, D. (trans. and ed.) (2000) *Libellus de Exordio Atque Procursu Istius, hoc est Dunhelmensis, Ecclesie: tract on the origins and progress of this church of Durham*. Oxford: Clarendon.
Stevenson, J. (ed.) (1858) *Chronicon monasterii de Abingdon*, 2 vols, Rolls Series. London: Longman, Brown, Green, Longmans and Roberts, 1.
Stubbs, W. (ed.) (1874) *Memorials of Saint Dunstan Archbishop of Canterbury*. London: Longman.
Teague, E. and V. Sankaran (trans.) (1978) '13: Buckinghamshire', in *Domesday Book*, ed. J. Morris. Chichester: Phillimore.
Thorn, C. and F. Thorn (trans.) (1979) '6: Wiltshire', in *Domesday Book*, ed. J. Morris. Chichester: Phillimore.
Whitelock, D. (ed. and trans.) (1930) *Anglo-Saxon Wills*. Cambridge: Cambridge University Press.
Winterbottom, M. and R.M. Thomson (eds) (2002) *William of Malmesbury Saints' Lives: Lives of SS. Wulfstan, Dunstan, Patrick, Benignus and Indract*. Oxford: Clarendon Press.

Secondary sources

Alcock, L. (1992) 'Message from the Dark Side of the Moon: western and northern Britain in the age of Sutton Hoo', in *The Age of Sutton Hoo: the seventh century in north-west Europe*, ed. M.O.H. Carver. Woodbridge: Boydell, 205–215.
Andersson Strand, E. (2014) 'Northerners: global travellers in the Viking Age', in *Global Textile Encounters*, eds M.-L. Nosch, Z. Feng and L. Varadarajan. Oxford: Oxbow Books, 75–80.
Anheuser, K. and M. Roumeliotou (2003) 'Characterisation of Mineralised Archaeological Textile Fibres through Chemical Staining', *The Conservator*, 27, 23–33.
Anon (2014) *Standard and Guidance for the Collection, Documentation, Conservation and Research of Archaeological Materials*. Reading: The Chartered Institute for Archaeologists.
Anon (2015) 'Piecing Together the Staffordshire Hoard', *Current Archaeology*, 305, 6–7.
Anon (2016) 'Britain in Archaeology', *British Archaeology*, 149, 10.
Anon (2016) '150 Anglo-Saxon graves found at Bulford', *Current Archaeology*, 315, 6–7.
Anon (2016) 'More Saxon graves on Salisbury Plan', *Current Archaeology*, 316, 8–9.
Arnold, J.H. (2008) *What is Medieval History?* Cambridge: Polity Press.
Bailey, R.N. (1989) 'St Cuthbert's Relics: some neglected Evidence', in *St Cuthbert, his Cult and his Community to AD 1200*, eds G. Bonner, D. Rollason and C. Stancliffe. Woodbridge: Boydell Press (repr. 2002), 231–246.
Baldwin Brown, G. and [Mrs.] A. Christie (1913) 'S. Cuthbert's Stole and Maniple at Durham', *The Burlington Magazine*, 23 (April 1913), 2–7, 9–11, 17.
Baldwin Brown, G. and [Mrs.] A. Christie (1913) 'S. Cuthbert's Stole and Maniple at Durham', *The Burlington Magazine*, 23 (May 1913), 67–69, 71–72.
Barlow, F., M. Biddle, O. von Feilitzen and D.J. Keene (1976) *Winchester in the Early Middle Ages: an edition and discussion of the Winton Domesday*. Oxford: Oxford University Press.
Baxter, S. and J. Blair (2006) 'Land Tenure and Royal Patronage in the Early English Kingdom: a model and a case study', *Anglo-Norman Studies*, 28, 19–46.
Bayless, A., J. Hines and K. Høilund Nielsen (2013) 'Interpretative Chronologies for the Female Graves', in *Anglo-Saxon Graves and Grave Goods of the 6th and 7th Centuries AD: A Chronological*

Framework, eds J. Hines and A. Bayless, The Society for Medieval Archaeology Monograph 33. London: The Society for Medieval Archaeology, 339–458.

Becker, C.J. (1953) 'Zwei Frauengräber des 7. Jahrhunderts aus Nørre Sandegaard, Bornholm', *Acta Archaeologica*, 24, 127–155.

Becker, C.J. (1990) *Nørre Sandegaard. Arkæologiske undersøkelser på Bornholm 1948-1952*, Historisk-filosofiske Skrifter 13. Copenhagen: The Royal Academy of Sciences and Letters.

Bédat, I. and B. Girault-Kurtzeman (2004) 'The Technical Study of the Bayeux Embroidery', in *The Bayeux Tapestry: Embroidering the Facts of History*, eds P. Bouet, B. Levy and F. Neveux. Caen: Presses Universitaires de Caen, 83–110.

Bennett, D.H. (1973) 'Seventh Century Cemeteries from the Ouse Valley', *Bedfordshire Archaeological Journal*, 8, 99–108,

Biddle, M. and B. Kjølbye-Biddle (2001) 'Repton and the Great Heathen Army, 873–4', in *Vikings and the Danelaw*, eds J. Graham-Campbell, R. Hall, J. Jesch and D.N. Parsons. Oxford: Oxbow Books, 45–96.

Blair, J. (2005) *The Church in Anglo-Axon England*. Oxford: Oxford University Press.

Bonde, N. and A.E. Christensen (1991/1992) *Dendrokronologisk datering af tømmer fra gravkamrene i Oseburg, Gokstad og Tune*. Oslo: Universitets Oldsaksamling Årbok, 153–160.

Bowsher, D., T. Dyson, N. Holder and I. Howell (eds) (2007) *The London Guildhall: an archaeological history of a neighbourhood from early medieval to modern times part 1*. London: Museum of London Archaeology.

Bradley, J. (1993) 'Moynagh Lough: an Insular Workshop of the Second or Third Quarter of the Eighth Century', in *The Age of Migrating Ideas*, eds R. Michael Spearman and J. Higgitt. Edinburgh: National Museums of Scotland, 74–81

Brandenburgh, C. (2012) 'Old Finds Rediscovered: two early medieval headdresses from the National Museum of Antiquities, Leiden, the Netherlands', *Medieval Dress and Textiles*, 8, 25–48.

Brown, S.A. (2013) *The Bayeux Tapestry, a sourcebook*. Turnhout: Brepols.

Bruce-Mitford, R. (1975) *The Sutton Hoo Ship-Burial volume 1: excavations, background, the ship, dating and inventory*. London: British Museum Publications.

Bruce-Mitford, R. (1983) *The Sutton Hoo Ship-Burial volume 3: late Roman and Byzantine silver hanging-bowls, drinking vessels, cauldrons and other containers, textiles, the lyre, pottery bottles and other items*. London: British Museum Publications.

Budny, M. (1984) 'The Anglo-Saxon Embroideries at Maaseik', *Klasse der Schone Kunsten*, 45, 57–133.

Budny, M. (1991) 'The Byrhtnoth Tapestry or Embroidery', in *The Battle of Maldon AD 991*, ed. D. Scragg. Oxford: Blackwell, 263–278.

Budny, M. and D. Tweddle (1984) 'The Maaseik Embroideries', *Anglo-Saxon England*, 13, 65–96.

Budny, M. and D. Tweddle (1985) 'The Early Medieval Textiles at Maaseik, Belgium', *The Antiquaries Journal*, 65, 353–389.

Butler, A. (1982) *The Batsford Encyclopaedia of Embroidery Stitches*, revised edition. London: Batsford.

Calberg, M. (1951) 'Tissus et Broderies Attribués aux Saintes Harlinde et Relinde', *Bulletin: Société Royale d'Archéologie de Bruxelles*, 1–26.

Campbell, E. and A. Lane (1989) 'Llangorse: a 10th-century royal crannog in Wales', *Antiquity*, 63, 675–681.

Carson Paston, E. and S.D. White with K. Gilbert (2014) *The Bayeux Tapestry and its Contexts: a reassessment*. Woodbridge: Boydell.

Chadwick Hawkes, S. (1982) 'The Archaeology of Conversions: Cemeteries', in *The Anglo-Saxons*, ed. J. Campbell. London: Phaidon; repr. London: Pengiun, 1991, 48–49.

Chadwick Hawkes, S. with G. Grainger (2003) *The Anglo-Saxon Cemetery at Worthy Park, Kingsworthy Near Winchester, Hampshire*, Oxford University School of Archaeology Monograph 59. Oxford: Oxford University School of Archaeology.

Christensen, A.E. (2006) 'Introduction', in *Osebergfunnet: Bind IV Tekstilene*, eds A.E. Christensen and M. Nöckert. Oslo: Museum of Cultural History, 353–354.

Christensen, A.E. and M. Nöckert (eds) (2006) *Osebergfunnet: Bind IV Tekstilene*. Oslo: Museum of Cultural History.

Christie, [Mrs] A.G.I. (1938) *English Medieval Embroidery: a brief survey of English embroidery dating from the beginning of the tenth century until the end of the fourteenth*. Oxford: Clarendon Press.

Clegg Hyer, M. and G.R. Owen-Crocker (2011) 'Making and Using Texiles', in *The Material Culture of Daily Living in the Anglo-Saxon World*, eds M. Clegg Hyer and G.R. Owen-Crocker. Exeter: University of Exeter Press, 157–184.

Cluckie, L. (2008) *The Rise and Fall of Art Needlework*. Bury St Edmunds: Arena Books.

Coatsworth, E. (2001) 'The Embroideries from the Tomb of St Cuthbert', in *Edward the Elder 899–924*, eds N.J. Higham and D. Hill. London: Routledge, 292–306.

Coatsworth, E. (2005) 'Stitches in Time: establishing a history of Anglo-Saxon embroidery', in *Medieval Clothing and Textiles*, 1, 1–27.

Coatsworth, E. (2014) '"A formidable undertaking": Mrs A.G.I. Christie and *English Medieval Embroidery*', in *Medieval Clothing and Textiles*, 10, 165–193.

Coatsworth, E. (2016) 'Opus What? The Textual History of Medieval Embroidery Terms and Their Relationship to the Surviving Embroideries', in *Textiles, Text, Intertext: essays in honour of Gale R. Owen-Crocker*, eds M. Clegg Hyer and J. Frederick. Woodbridge: Boydell Press, 43–67.

Coatsworth, E. and G.R. Owen-Crocker (2007) *Medieval Textiles of the British Isles AD 450-1100*. Oxford: Archaeopress.

Coatsworth, E. and M. Pinder (2002) *The Art of the Anglo-Saxon Goldsmith*. Woodbridge: Boydell Press (repr. 2010).

Coldicott, D.K. (1989) *Hampshire Nunneries*. Chichester: Phillimore.

Cole, A.S. (ed. and trans.) (1888) Ernest Lefébure, *Embroidery and Lace*. London: H. Grevel and Co.

Cooke, B. (1990) 'Fibre Damage in Archaeological Textiles', in *Archaeological Textiles: occasional papers 10*, eds S.A. O'Conner and M.M. Brooks. London: The United Kingdom Institute for Conservation, 5–14.

Cooke, B. and B. Lomas (1990) 'The Evidence of wear and damage in ancient textiles', in *Textiles in Northern Archaeology: NESAT III: textile symposium in York 6-9 May 1987*, eds P. Walton and J.-P. Wild. London: Archetype, 215–226.

Cramp, R. (2006) *Corpus of Anglo-Saxon Stone Sculpture volume 7: South-West England*. Oxford: Oxford University Press.

Cronyn, J.M. (1990) *The Elements of Archaeological Conservation*. London: Routledge (repr. 2001).

Crook, J. (2011) *English Medieval Shrines*. Woodbridge: Boydell.

Crowfoot, E.G. (1956) 'Appendix I. Objects found in Cremation Heath, Mound II', in M. Posnansky, 'The Pagan Barrow Cemetery at Heath Wood, Ingleby: 1955 excavations', *Journal of the Derbyshire Archaeological and Natural History Society*, 76, 40–56.

Crowfoot, E. (1983) 'The Textiles', in *The Sutton Hoo Ship-Burial volume 3*, eds R. Bruce-Mitford and A.C. Evans. London: British Museum Publications, 404–479.

Crowfoot, E. (1987) 'The Textiles', in *Dover: the Buckland Anglo-Saxon cemetery*, ed. Vera I. Evison, Historic Buildings and Monuments Commission for England Archaeological Report No. 3. London: Historic Buildings and Monuments Commission for England, 190–195.

Crowfoot, E. (1990) 'Personal Possessions: textiles', in *Object and Economy in Medieval Winchester*, ed. M. Biddle, Winchester Studies, 7.ii. Oxford: Clarendon Press, 467–493.

Crowfoot, E. (1990) 'Textile Fragments from "Relic-Boxes" in Anglo-Saxon Graves', in *Textiles in Northern Archaeology: NESAT III: textile symposium in York, 6-9 May 1987*, eds P. Walton and J.-P. Wild. London: Archetype Publications, 47–56.

Crowfoot, E. (2003) 'Chapter 5: The Textile Remains', in *The Anglo-Saxon Cemetery at Worthy Park, Kingsworthy Near Winchester, Hampshire*, Oxford University School of Archaeology Monograph 59. Oxford: Oxford University School of Archaeology, 192–195.

Crowfoot, G.M. (1939) 'The Tablet-woven Braids from the Vestments of St Cuthbert at Durham', *The Antiquaries Journal*, 19, 57–80.

Crowfoot, G.M. (1952) 'Anglo-Saxon Tablet Weaving', *The Antiquaries Journal*, 32, 189–191.

Crowfoot, G.M. (1956) 'Note on a Fragment of Embroidery from the Basilica Ambrosiana in Milan', in *The Relics of St Cuthbert*, ed. C.F. Battiscombe. Oxford: Oxford University Press, 392–394.

Crowfoot, G.M. (1956) 'The Braids', in *The Relics of St Cuthbert*, ed. C.F. Battiscombe. Oxford: Oxford University Press, 433–469.

Cunliffe, B. (2013) *Britain Begins*. Oxford: Oxford University Press.

d'Arzago, A. de Capitani (1941) *Antichi Tessuti della Basilica Ambrosiana*, new series 2. Milan: Biblioteca de L'Arte.

de Farcy, M.L. (1890–1919) *La Broderie de XI$^{\grave{e}rrie}$ siècle jusqu'à nos jours d'après des spécimens authentiques et les anciens inventaires*, 3 vols. Angers: Belhomme, Libraire-Éditeur.

Dodwell, C.R. (1982) *Anglo-Saxon Art: a new perspective*. Manchester: Manchester University Press.

Dyer, C. (2009) *Making a Living in the Middle Ages: the people of Britain 850-1520*. London: Yale University Press.

Edwards, G. (1991) '302. MLK 76 [1053] <543>, Pit 55', in *Aspects of Saxo-Norman London: II Finds and Environmental Evidence*, ed. A. Vince. London: London and Middlesex Archaeological Society.

East, K. (1983) 'The Shoes', in *The Sutton Hoo Ship-Burial volume 3*, eds R. Bruce-Mitford and A.C. Evans. London: British Museum Publications, 788–812.

Egan, G. (2007) '8.8 Accessioned Finds', in *The London Guildhall: an archaeological history of a neighbourhood from early medieval to modern times part II*, eds D. Bowsher, T. Dyson, N. Holder and I. Howell, MOLA Monograph 36. London: Museum of London Archaeology, 446–472.

Emery, I. (1980) *The Primary Structures of Fabrics*. London: Thames and Hudson (repr. 2009).

Evans, A. (2005) 'Seventh-Century Assemblages', in *Sutton Hoo: an Anglo-Saxon princely burial ground and its context*, ed. M. Carver, Reports of the Research Committees of the Society of Antiquaries of London, 69. London: British Museum Press, 201–282.

Fell, C. (1984) *Women in Anglo-Saxon England*. London: British Museum Publications.

Findlay, M. (1984) 'Report on the Conservation of the Orkney Hood', in *The Laboratories of the National Museum of Antiquities of Scotland 2*, eds T. Bryce and J. Tate. Edinburgh: National Museum of Antiquities of Scotland, 95–102.

Fitch, [Rev.] S. (1863–1864) 'Discovery of Saxon Remains at Kempston', in *Reports and Papers read at The Meetings of the Architectural Societies of the County of York, Diocese of Lincoln, Archdeaconry of Northampton, County of Bedford, Diocese of Worcester and County of Leicester*. Lincoln: Brookes and Vibert, 269–299.

Foot, S. (2000) *Veiled Women vol. 1: the disappearance of nuns from Anglo-Saxon England*, 2 vols. Aldershot: Ashgate.

Foot, S. (2006) *Monastic Life in Anglo-Saxon England c. 600-900*. Cambridge: Cambridge University Press.

Foreman, M. (1998) 'Work-Boxes (Graves I, II, 183)', in *The Anglo-Saxon Cemetery at Castledyke South, Barton-on-Humber*, eds G. Drinkall and M. Foreman. Sheffield: Sheffield Academic Press, 285.

Freeman, C. (2011) *Holy Bones, Holy Dust: How Relics Shaped the History of Medieval Europe*. London: Yale University Press.

Freyhan, R. (1956) 'The Stole and Maniples (c) the place of the stole and maniple in Anglo-Saxon art of the tenth century', in *The Relics of St Cuthbert*, ed. C.F. Battiscombe. Oxford: Oxford University Press, 409–432.

Gabra-Sanders, T. (2001) 'The Orkney Hood, Re-Dated and Re-Considered', in *The Roman Textile Industry and its Influence: a birthday tribute to John Peter Wild*, eds P. Walton Rogers, L. Bender Jørgensen and A. Rast-Eicher. Oxford: Oxbow Books, 99–104.

Garside, P. (2012) 'Gold and Silver Metal Thread', in *Encyclopedia of Medieval Dress and Textiles of the British Isles c. 450-1450*, eds G. Owen-Crocker, E. Coatsworth and M. Hayward. Leiden: Brill, 237–239.

Geake, H. (1997) *The Use of Grave Goods in Conversion-Period England, c. 600-c. 850*, British Archaeological Reports, British Series 261. Oxford: Archaeopress.

Geijer, A. (1938) *Birka III: Die Textilfunde aus den Grabern*. Uppsala: Kungl Vitterhets Historie de Antikvitets Akademien.

Geijer, A. (1979) 'The Textile finds from Birka', *Acta Archaeologica*, 50, 209–222.

Gell, A. (1999) *The Art of Anthropology: essays and diagrams*, ed. E. Hirsch. London: Athlone Press.

Gillard, R.D., S.M. Hardman, R.G. Thomas and D.E. Watkinson (1994) 'The Mineralization of Fibres in Burial Environments', *Studies in Conservation*, 39, 132–140.

Gillis, C. and M.-L. B. Nosch (eds) (2007) *First Aid for the Excavation of Archaeological Textiles*. Oxford: Oxbow Books.

Glover, J.M. (1992) 'Conservation and Storage: textiles', in *Manual of Curatorship: a guide to museum practice*, Second edition, ed. J.M.A. Thompson. Oxford: Butterworth-Heinemann (repr, 1994), 302–339.

Goodway, M. (1987) 'Fiber Identification in Practice', *Journal of the American Institute for Conservation*, 26, 27–44.

Graham-Campbell, J.A. (2002) 'Tenth-Century Graves: The Viking-Age Artefacts and their Significance', in *Excavation on St Patrick's Isle Peel, Isle of Man, 1982-88 Prehistoric, Viking, Medieval and Later*, eds A.M. Cubbon, P.J. Davey and M. Gelling. Liverpool: Liverpool University Press, 83–98.

Granger-Taylor, H. (1983) 'The Two Dalmatics of Saint Ambrose?', *Bulletin de Liaison du Centre International D'Ètude des Textiles Anciens*, 57-58, 127–173.

Granger-Taylor, H. (1989) 'The Weft-Patterned Silks and their Braid: the remains of an Anglo-Saxon dalmatic of *c*. 800?', in *St Cuthbert, his Cult and his Community to AD 1200*, eds G. Bonner, D. Rollason and C. Stancliffe. Woodbridge: Boydell Press (repr. 2002), 303–327.

Granger-Taylor, H. and F. Pritchard (2001) 'A Fine Quality Insular Embroidery from Llan-gors Crannóg, near Brecon', in *Pattern and Purpose in Insular Art: proceedings of the fourth international conference on insular art held at the National Museum and Gallery, Cardiff 3-6 September 1998*, eds M. Redknap, N. Edwards, S. Youngs, A. Lane and J. Knight. Oxford: Oxbow, 91–99.

Green, V. (1796) *The History and Antiquities of the City and Suburbs of Worcester*, 2 vols. London: W. Bulmer & co., vol. I.

Gretsch, M. (2001) 'The Junius Psalter Gloss', in *Edward the Elder 899-924*, eds N.J. Higham and D.H. Hill. London: Routledge, 280–291.

Griffith, A.F. and L.F. Salzmann (1914) 'An Anglo-Saxon Cemetery at Alfriston, Sussex', *Sussex Archaeological Collections, Relating to the History and Antiquities of the County*, 56, 16–51.

Griffiths, D. (2011) 'The Ending of Anglo-Saxon England: identity, allegiance, and nationality', in *The Oxford Handbook of Anglo-Saxon Archaeology*, eds H. Hamerow, D.A. Hinton and S. Crawford. Oxford: Oxford University, 62–78.

Guðjónsson, E.E. (1985) *Traditional Icelandic Embroidery*. Reykjavik; Iceland Review.

Hald, M. (1980) *Ancient Danish Textiles from Bogs and Burials: a comparative study of costume and Iron Age textiles*, Archaeological-Historical Series 21. Copenhagen: National Museum of Denmark.

Hald, M.M., P. Henriksen, L. Jørgensen and I. Skals (2015) 'Danmarks ældste løg', in *Nationalmusets Arbejdsmark 2015*, eds M.M. Hald, P. Henriksen, L. Jørgensen and I. Skals. Copenhagen: National Museum of Denmark, 104–115.

Hall, R.A. (1998) 'A Silver Appliqué from St Mary Bishophill Senior, York', *The Yorkshire Archaeological Journal*, 70, 61–66.

Hall, R.A. (2014) *Anglo-Scandinavian Occupation at 16-22 Coppergate: defining a townscape*, The Archaeology of York Anglo-Saxon York 8/5. York: Council for British Archaeology.

Hall, R.A., D.W. Rollason, M. Blackburn, D.N. Parsons, G. Fellows-Jensen, A.R. Hall, H.K. Kenward, T.P. O'Connor, D. Tweddle, A.J. Mainman and N.S.H. Rogers (2004) *Aspects of Anglo-Scandinavian York*, The Archaeology of York, Anglo-Scandinavian York 8/4. York: Council for British Archaeology.

Hamerow, H. (2012) *Rural Settlements and Society in Anglo-Saxon England*. Oxford: Oxford University Press.

Hedges, R.E.M., R.A. Housley, C. Bronk-Ramsey and G.J. van Klinken (1993) 'Radiocarbon Dates from the Oxford AMS System: Archaeometry datelist 16', *Archaeometry*, 35, 147–167.

Henman, C. (1871) 'Fragments of Ancient Vestments', in *Architectural Association Sketchbook*, 16 vols. London: Architectural Association, 1867–1913, IV.

Henshall, A.S. (1951–2) 'Early Textiles Found in Scotland', *Proceedings of the Society of Antiquaries of Scotland*, 86, 1–29.

Herlihy, D. (1990) *Opera Muliebria: women and work in Medieval Europe*. London: McGraw-Hill.

Higham, N.J. (2013) 'From Tribal Chieftains to Christian Kings', in *The Anglo-Saxon World*, eds N.J. Higham and M.J. Ryan. London: Yale University Press, 126–178.

Higham, N.J. and M.J. Ryan (eds) (2013) *The Anglo-Saxon World*. London: Yale University Press.

Hines, J. (1993) *Clasps Hektespenner Agraffen: Anglo-Scandinavian Clasp of Classes A-C of the 3rd to 6th centuries AD. Typology, Diffusion and Function*. Uddevalla: Bohusläningens Boktryckeri AB.

Hinton, D.A. (1996) *Southampton Finds volume 2: the gold, silver and other non-ferrous alloy objects from Hamwic*. Stroud: Alan Sutton Publishing, 48–50.

Hinton, D.A. (2005) *Gold and Gilt, Pots and Pins: possessions and people in Medieval Britain*. Oxford: Oxford University Press.

Hohler, C. (1956) 'The Stole and Maniples (b) the iconography', in *The Relics of St Cuthbert*, ed. C.F. Battiscombe. Oxford: Oxford University Press, 396–408.

Hollis, S. (2004) 'Wilton as a Centre of Learning', in *Writing the Wilton Women: Goscelin's Legend of Edith and Liber confortatorius*, eds S. Hollis with W.R. Barnes, R. Hayward, K. Loncar and M. Wright. Turnhout: Brepols, 307–338.

Holtorf, C. (2002) 'Notes on the Life of a Pot Sherd', *Journal of Material Culture*, 7, 49–71.

Horie, C.V. (1992) 'Conservation and Storage: leather objects', in *Manual of Curatorship: a guide to museum practice*, Second edition, ed. John M.A. Thompson. Oxford: Butterworth-Heinemann (repr, 1994), 340–345.

Howlett, D.R., J. Blundell, T. Christchev and C. White (2001) *Dictionary of Medieval Latin from British Sources: fascicule VI M*. Oxford: Oxford University Press.

Hulse, L. (2014) 'Elizabeth Burden and The Royal School of Needlework', *The Journal of William Morris Studies*, 22, 22–34.

Hulse, L. (2017) '"When Needlework was at its very finest": Opus Anglicanum and its influence on May Morris', in *May Morris: Art and Life. New Perspectives*, ed. L. Hulse. London: Friends of William Morris Gallery, 87–110.

Ingold, T. (2011) *Being Alive: essays on Movement, Knowledge and Description*. London: Routledge.

Ingstad, A.S. (1993) 'Tekstilene i Osebergskipet', in *Oseberg - Dronningens Grav: vår arkeologiske nasjonalskatt i nytt lys*, eds A.E. Christensen, A.S. Ingstad an B. Myhre. Oslo: Schibsted, 176–208.

Ivy, J. (1997) *Embroideries at Durham Cathedral*, Second edition. Sunderland: Attey and Sons.

Johnstone, P. (2002) *High Fashion in the Church*. Leeds: Maney.

Jones, A. (2002) *Archaeological Theory and Scientific Practice*. Cambridge: Cambridge University Press.

Jones, J., J. Unruh, R. Knaller, I. Skals, L. Ræder Knudsen, E. Jordan-Fahrbach and L. Mumford (2007) 'Guidelines for the Excavation of Archaeological Textiles', in *First Aid for the Excavation of Archaeological Textiles*, eds C. Gillis and M.-L.B. Nosch. Oxford: Oxbow Books, 5–30.

Joy, J. (2009) 'Reinvigorating Object Biography: reproducing the drama of object lives', *World Archaeology*, 41, 540–556.

Kendrick, A.F. (1905) *English Embroidery*. London: George Newnes.

Kennett, D.H. (1986) 'Recent work on the Anglo-Saxon Cemetery found at Kempston', *South Midlands Archaeology* 16, 3–14.

Keynes, S. (2014) 'Cnut', in *The Wiley Blackwell Encyclopedia of Anglo-Saxon England*, Second edition, eds M. Lapidge, J. Blair, S. Keynes and D. Scragg. Oxford: Wiley Blackwell, 111–112.

King, B.M. (2019) *The Wardle Family and it Circle*. Woodbridge: Boydell Press.

Knappett, C. (2005) *Thinking through Material Culture: an interdisciplinary perspective*. Philadelphia: University of Pennsylvania Press.

Komatsu, M. (2007) 'Investigation of the Fragments of Liturgical Textiles from Worcester Cathedral'. Unpublished MA dissertation, University of Southampton, Textile Conservation Centre.

Kopytoff, I. (1986) 'The Cultural Biography of things: commoditization as process', in *The Social Life of Things: commodities in cultural perspective*, ed. A. Appadurai. Cambridge: Cambridge University Press (repr. 2011), 64–91.

Landi, S. (1998) *The Textile Conservator's Manual*, Second edition. London: Routledge.

Laporte, J.-P. and R. Boyer (1991) *Trésors de Chelles*: sépultures et reliques de la Reine Bathilde et de l'Abbesse Bertille. Chelles: Societé Archéologique et Historique de Chelles.

Larratt Keefer, S. (2007) 'A Matter of Style: clerical vestments in the Anglo-Saxon church', *Medieval Clothing and Textiles*, 3, 13–39.

Latham, R.E., D.R. Howlett, and R.K. Ashdowne (eds) (1975–2013) *Dictionary of Medieval Latin from British Sources*. Oxford: Oxford University Press, fascicule I A-B, II C, V I-L, IX P-Pel.

Leahy, K. and R. Bland (2009) *The Staffordshire Hoard*. London: British Museum Press (repr. 2010).

Lester-Makin, A. (2012) 'Stitches' and 'Stitches: filling stitches and other embroidery techniques', in *Encyclopedia of Medieval Dress and Textiles of the British Isles c. 450-1450*, eds G.R. Owen-Crocker, E. Coatsworth and M. Hayward. Leiden: Brill, 547–550 and 550–564.

Lester-Makin, A. (2016) 'The Front Tells the Story, the Back Tells the History: a technical discussion of the embroidering of the Bayeux Tapestry', in *Making Sense of the Bayeux Tapestry: Readings and Reworkings*, eds A.C. Henderson with G.R. Owen-Crocker. Manchester: Manchester University Press, 23–40.

Lester-Makin, A. (2017) 'Looped Stitch: the travels and development of an embroidery stitch', in *The Daily Lives of the Anglo-Saxons*, eds C. Biggam, C. Hough and D. Izdebska, Essays in Anglo-Saxon Studies 8. Tempe Arizona: Arizona Center for medieval and Renaissance Studies, 119–136.

Lester-Makin, A. (2018) 'Les six châteaux de la Tapisserie de Bayeux: Une discussion technique du travail de broderie de la Tapisserie de Bayeux', in *L'Invention de la Tapisserie de Bayeux: naissance, composition et style d'un chef-d'œuvre médiéval*, eds S. Lemagnen, S.A. Brown and G. Owen-Crocker. Rouen: Point de Vues and Musée de la Tapisserie de Bayeux, 73–91.

Leyser, H. (1995) *Medieval Women: a social history of women in England 450-1500*. London: Weidenfeld & Nicolson (repr. London: Phoenix, 1997).

Love, R.C. (2014) 'Æthelthryth', in *The Wiley Blackwell Encyclopaedia of Anglo-Saxon England*, Second edition, eds M. Lapidge, J. Blair, S. Keynes and D. Scragg. Chichester: John Wiley & Sons, 19–20.

MacConnoran, P. (2011) 'Complete Catalogue of Selected Medieval Fragments from 1 Poultry' (on accompanying CD) in *The Development of Early Medieval and Later Poultry and Cheapside*, ed. A. Marriott, MOLA Monograph 38. London: Museum of London Archaeology.

MacConnoran, P. with A. Nailer (2007) 'Complete Catalogue of Leather Items' (on accompanying CD), in *The London Guildhall: an archaeological history of a neighbourhood from early medieval to modern times part II*, eds D. Bowsher, T. Dyson, N. Holder and I. Howell, MOLA Monograph 36. London: Museum of London Archaeology.

MacConnoran, P. with A. Nailer (2007) '8.15 Leather Items', in *The London Guildhall: an archaeological history of a neighbourhood from early medieval to modern times part II*, eds D. Bowsher, T. Dyson, N. Holder and I. Howell, MOLA Monograph 36. London: Museum of London Archaeology, 479–485.

Marshall, F. and H. Marshall (1894) *Old English Embroidery*. London: Horace Cox.

Mayr-Harting, H. (2011) *Religion, Politics and Society in Britain 1066-1272*. Harlow: Pearson Education.

Meaney, A.L. (1981) *Anglo-Saxon Amulets and Curing Stones*, British Archaeological Reports, British Series, 96. Oxford: Archaeopress.

Messent, J. (2010) *The Bayeux Tapestry Embroiderers' Story*, Second edition. Tunbridge Wells: Search Press.

Miller, M.C. (2014) *Clothing the Clergy: virtue and power in Medieval Europe c. 800-1200*. London: Cornell University Press.

Moreland, J. (1999) 'The World(s) of the Cross', *World Archaeology*, 31, 194–213.

Moreland, J. (2010) *Archaeology, Theory and the Middle Ages*. London: Duckworth.

Morrell, A. (1999) *Guide to Structural Sewing: terms and techniques*. Ahmedabad: Sarabhai Foundation.
Morrell, A. (2007) *The Migration of Stitches and the Practice of Stitches as Movement*. Ahmedabad: D.S. Mehta.
Morris, M. (1893) *Decorative Needlework*. London: Joseph Hughes.
Morton, A.D. (1992) *Excavations at Hamwic: Vol. 1 Excavations 1946-83, excluding Six Dials and Melbourne Street*, CBA Research Report 84. London: Council for British Archaeology.
Mostert, M. (2014) 'Willibrord', in *The Wiley Blackwell Encyclopedia of Anglo-Saxon England*, Second edition, eds M. Lapidge, J. Blair, S. Keynes and D. Scragg. Oxford: Wiley Blackwell, 499.
Mould, Q., I. Carlisle and E. Cameron (eds) (2003) *Leather and Leatherworking in Anglo-Scandinavian and Medieval York*, The Archaeology of York The Small Finds 17/16. York: Council for British Archaeology.
Mould, Q., I. Carlisle and E. Cameron (2003) 'Catalogue', in *Leather and Leatherworking in Anglo-Scandinavian and Medieval York*, The Archaeology of York The Small Finds 17/16, eds Q. Mould, I. Carlisle and E. Cameron. York: Council for British Archaeology, 3439–3529.
Mould, Q., I. Carlisle and E. Cameron (2003) 'Everyday Life', in *Leather and Leatherworking in Anglo-Scandinavian and Medieval York*, The Archaeology of York The Small Finds 17/16, eds Quita Mould, Ian Carlisle and Esther Cameron. York: Council for British Archaeology, 3268–3417.
Mumford, L. (2002) 'The Conservation of the Llangorse Textile', in *Proceedings of the 8th ICOM Group on Wet Organic Archaeological Materials Conference*, eds P. Hoffmann, J.A. Spriggs, T. Grant, C. Cook and A. Recht. Bremerhaven: International Council of Museums (ICOM), 471–491.
Mumford, L., H. Prosser and J. Taylor (2007) 'The Llangorse Textile: approaches to understanding an early medieval masterpiece', in *Ancient Textiles: production, craft and society*, eds C. Gillis and M.-L.B. Nosch. Oxford: Oxbow Books, 158–162.
Nailer, A. and P. Reid with P. MacConnoran (2011) 'Leather and Shoes', in *The Development of Early Medieval and Later Poultry and Cheapside Excavations at 1 Poultry and Vicinity, City of London*, eds M. Burch and P. Treveil with D. Keene. London: Museum of London Archaeology, 332–341.
Naylor, J. (2004) *An Archaeology of Trade in Middle Saxon England*, British Archaeological Reports, British Series 376. Oxford: Archaeopress.
Nöckert, M. (2002) 'The Oseberg Textiles and the Orient', *Bulletin du CIETA*, 79, 44–50.
O'Sullivan, A., F. McCormick, T.R. Kerr, L. Harney and J. Kinsella (2014) *Early Medieval Dwellings and Settlements in Ireland, AD 400-1100*, British Archaeological Reports, International Series 2604. Oxford: Archaeopress.
Ottaway, P. and C.A. Morris (2003) 'The Leatherworking Tools Recovered' in *Leather and Leatherworking in Anglo-Scandinavian and Medieval York*, The Archaeology of York The Small Finds 17/16, eds Q. Mould, I. Carlisle and E. Cameron. York: Council for British Archaeology, 3235–3243.
Owen-Crocker, G.R. (2004) *Dress in Anglo-Saxon England, revised and enlarged*. Woodbridge: Boydell Press.
Owen-Crocker, G.R., E. Coatsworth and M. Hayward (2012) *Encyclopedia of Medieval Dress and Textiles of the British Isles c. 450-1450*. Leiden: Brill.
Perkins, M. (2007) 'A Stole, Maniple and Four other Embroideries in Worcester Cathedral Library', *Archaeology at Worcester Cathedral: report of the sixteenth annual symposium March 2006*, 3–12.
Plenderleith, E. (1956) 'The Stole and Maniples (a) the technique', in *The Relics of St Cuthbert*, ed. C.F. Battiscombe. Oxford: Oxford University Press, 375–396.
Plunkett, S.J. (1999) 'The Anglo-Saxon Loom from Pakenham, Suffolk', *Proceedings of the Suffolk Institute of Archaeology and History*, 34, 277–298.
Posnansky, M. (1956) 'The Pagan-Danish Barrow Cemetery at Heath Wood, Ingleby: 1955 excavations', *Journal of the Derbyshire Archaeological and Natural History Society*, 76, 40–56.
Powlesland, D. (1997) 'Early Anglo-Saxon Settlements, Structures, Form and Layout, in *The Anglo-Saxons from the Migration Period to the Eighth Century: an ethnographic perspective*, ed. J. Hines. Woodbridge: Boydell, 101–124.

Pritchard, F. (1991) 'Leather Work', in *Aspects of Saxo-Norman London: II Finds and Environmental Evidence*, ed. A. Vince. London: London and Middlesex Archaeological Society, 211–240.

Pritchard, F. (1988) 'Decoration', in *Shoes and Pattens*, eds F. Grew and M. de Neergaard. London: Museum of London (repr. Woodbridge: Boydell and Brewer, 2006), 75–90.

Pritchard, F. (2014) 'Textiles from Dublin', in *Vikingatidens Kvinnor*, eds N. Coleman and N. Løkka. Oslo: Scandinavian Academic Press, 225–240.

Purcell, E. and J. Sheehan (2013) 'Viking Dublin: enmities, alliances and the cold gleam of silver', in *Everyday Life in Viking-Age Towns: Social Approaches to Towns in England and Ireland, c. 800-1100*, eds D.M. Hadley and L.T. Harkel. Oxford: Oxbow Books, 35–60.

Pye, E. (1992) 'Conservation and Storage: archaeological material', in *Manual of Curatorship: a guide to museum practice*, Second edition, ed. J.M.A. Thompson. Oxford: Butterworth-Heinemann (repr. 1994), 392–426.

Redknap, M. and A. Lane (1994) 'The Early Medieval Crannog at Llangorse, Powys: an interim statement of the 1989–1993 seasons', *International Journal of Nautical Archaeology*, 23, 189–205.

Richards, J.D. (1991) *Viking Age England*. London: B.T. Batsford/English Heritage (repr. Stroud: The History Press, 2010).

Richards, J.D. (1992) 'Anglo-Saxon Symbolism', in *The Age of Sutton Hoo: the seventh century in north-west Europe*, ed. M.O.H. Carver. Woodbridge: Boydell Press, 131–147.

Raine, J. (1828) *Remarks on the 'Saint Cuthbert' of the Rev. James Raine, M.A.* Newcastle: Preston & Heaton.

Rock, [Rev.] D. (1870) *South Kensington Museum Textile Fabrics*. London: Chapman and Hall.

Rock, [Rev.] D. (1876) *Textile Fabrics*. London: Chapman and Hall.

Ryan, M.J. (2013) 'The Age of Æthelred', in *The Anglo-Saxon World*, eds N.J. Higham and M.J. Ryan. London: Yale University Press, 335–373.

Ryder, M.L. (1983) 'Appendix 1: Wool from Anglo-Saxon Sites', in *The Sutton Hoo Ship-Burial volume 3: late Roman and Byzantine silver hanging-bowls, drinking vessels, cauldrons and other containers, textiles, the lyre, pottery bottles and other items*, eds R. Bruce-Mitford and A.C. Evans. London: British Museum Publications, 463–464.

Scull, C.J. (1992) 'Before Sutton Hoo: structures of power and society in early East Anglia', in *The Age of Sutton Hoo: the seventh century in north-west Europe*, ed. M.O.H. Carver. Woodbridge: Boydell Press, 3–22.

Scull, C. (1997) 'Urban Centres in Pre-Viking England?', in *The Anglo-Saxons from the Migration Period to the Eighth Century: an ethnographic perspective*, ed. J. Hines. Woodbridge: Boydell, 269–310.

Sharp, S. (2001) 'The West Saxon tradition of dynastic marriage: with special reference to Edward the Elder', in *Edward the Elder 899-924*, ed. N.J. Higham and D. Hill. London: Routledge, 79–88.

Sibley, L.R. and K.A. Jakes (1984) 'Survival of protein fibres in archaeological contexts', *Science and Archaeology*, 26, 17–27.

Smith, A. (1999) 'The Needle Tidy', in *Scar: A Viking Boat Burial on Sanday, Orkney*, eds O. Owen and M. Dalland. East Linton: Tuckwell Press in association with Historic Scotland, 95–96.

Southwell, [Canon] H.B. (1913) 'A Descriptive Account of some Fragments of Mediæval Embroidery found in Worcester Cathedral', *Reports and Papers of the Associated Architectural Societies*, 32, 150–165, plates I–XVI.

Speake, G. (1980) *Anglo-Saxon Animal Art and its Germanic Background*. Oxford: Oxford University Press.

Speed, G. (2014) *Towns in the Dark? Urban Transformations from Late Roman Britain to Anglo-Saxon England*. Oxford: Archaeopress.

Spriggs, J.A. (2003) 'Conservation of the Leatherwork', in *Leather and Leatherworking in Anglo-Scandinavian and Medieval York*, The Archaeology of York, The Small Finds 17/16, eds Q. Mould, I. Carlisle and E. Cameron. York: Council for British Archaeology, 3213–3221.

St John Hope, W.H. (1892) *Proceedings of the Society of Antiquaries*, 2nd ser., 14, 196–200.

Stafford, P. (2001) *Queen Emma and Queen Edith: queenship and women's power in eleventh-century England*, Second edition. Oxford: Wiley-Blackwell.

Staniland, K. (2007) *Medieval Craftsmen: Embroiderers,* Seventh edition. London: British Museum Press.

Stevens, H.M. (1990) 'Maaseik Reconstructed: a practical investigation and interpretation of 8th-century embroidery techniques', in *Textiles in Northern Archaeology: NESAT III,* eds P. Walton and J.P. Wild. London: Archetype Publications, 57–60.

Tester, A., S. Anderson, I. Riddler and R. Carr (2014) *Staunch Meadow, Brandon, Suffolk: a high status Middle Saxon settlement on the fen edge,* East Anglian Archaeology 151, ed. K. Wade. Bury St Edmunds: Suffolk County Council Archaeology Service.

Thomas, S. (1980) *Medieval Footwear from Coventry: a catalogue of the Collection of the Coventry Museums.* Coventry: Coventry Museums.

Thornton, J.H. (1990) 'Shoes, Boots, and Shoe Repairs', in *Object and Economy in Medieval Winchester,* ed. M. Biddle, Winchester Studies, 7.ii. Oxford: Clarendon Press, 591–617.

Tipper, J. (2004) *The Grubenhaus in Anglo-Saxon England: An analysis and interpretation of the evidence from a most distinctive building type.* Yedingham: Landscape Research Centre.

Tweddle, D. (1989) 'Comments on the use of the Pouch as a Reliquary', in *Textiles, Cordage and Raw Fibre from 16-22 Coppergate,* 17/5, ed. P. Walton. London: Council for British Archaeology, 378–381.

Ulmschneider, K. (2011) 'Settlement Hierarchy', in *The Oxford Handbook of Anglo-Saxon Archaeology,* eds H. Hamerow, D.A. Hinton and S. Crawford. Oxford: Oxford University Press, 156–171.

Vedeler, M. (2014) *Silk for the Vikings,* Ancient Textiles Series, 15. Oxford: Oxbow Books.

Vernon, C. (2018) 'Dressing for Succession in Norman Italy: the mantle of King Roger II', *Al-Masāq Journal of the Medieval Mediterranean,* DOI: https://www.tandfonline.com/doi/full/10.1080/09503110.2018.1551699.

de Villard, U.M. (1940) 'Una Iscrizione Marwanide su Stoffa del Secolo XI nella Basilica de S. Ambrogio a Milano', *Oriente Moderno,* 20 (October 1940), 504–506.

Vince, A. (1991) 'The Development of Saxon London', in *Aspects of Saxon-Norman London: II Finds and Environmental Evidence,* ed. A. Vince. London: London and Sussex Archaeological Society, 409–435.

Volken, M. (2014) *Archaeological Footwear: development of shoe patterns and styles from Prehistory till the 1600's.* Zwolle: Spa-Uitgevers.

Walton, P. (1989) *Textiles, Cordage and Raw Fibre from 16-22 Coppergate,* The Archaeology of York 17/5. London: Council for British Archaeology.

Walton Rogers, P. (1997) *Textile Production at 16-22 Coppergate,* The Archaeology of York 17/11. York: Council for British Archaeology.

Walton Rogers, P. (2005) 'Gold Thread', in *The Origins of Mid-Saxon Southampton: Excavations at the Friends Provident St Mary's Stadium 1998-2000,* eds V. Birbeck, R.J.C. Smith, P. Andrews and N. Stoodley. Salisbury: Wessex Archaeology, 68–69.

Walton Rogers, P. (2005) 'The Textiles from Mounds 5, 7, 14 and 17', in *Sutton Hoo: an Anglo-Saxon princely burial ground and its context,* ed. M. Carver. London: British Museum Press, 262–268.

Walton Rogers, P. (2007) *Cloth and Clothing in Early Anglo-Saxon England: AD 450-700.* York: Council for British Archaeology.

Walton Rogers, P. (2007) '6.3 The importance and organisation of textile production', in *Rural Settlement, Lifestyles and Social Change in the Later First Millennium AD: Anglo-Saxon Flixborough in its wider context,* Excavations at Flixborough vol. 4, ed. C. Loveluck. Oxford: Oxbow Books, 106–111.

Walton Rogers, P. (2009) 'Textile Production', in *Life and Economy at Early Medieval Flixborough c. AD 600-100: the artefact evidence,* Excavations at Flixborough vol. 2, ed. D.H. Evans and C. Lovekuck. Oxford: Oxbow Books, 281–316.

Walton Rogers, P. (2014) 'Cloth, Clothing and Anglo-Saxon Women', in *A Stitch in Time: Essays in Honour of Lise Bender Jørgensen,* eds S. Bergerbrant and S.H. Fossøy. Gothenburg: Gothenburg University, Department of Historical Studies, 253–280.

Walton Rogers, P. (2014) 'Textile Production and Treatment', in *Staunch Meadow, Brandon, Suffolk: a high status Middle Saxon settlement on the fen edge,* East Anglian Archaeology 151, ed. K. Wade. Bury St Edmunds: Suffolk County Council Archaeology Service, 285–294.

Walton Rogers, P. (2018) 'Textiles, Cords, Animal Fibres and Human Hair' in *28-9 High Ousegate, York, UK*, eds N. Macnab and J. McComish. York: York Archaeological Trust, 14–41.

Webster, L. (1991) 'Metalwork', in *The Making of England: Anglo-Saxon art and culture AD 600-900*, eds L. Webster and J. Backhouse. London: British Museum Press, 220–239.

Webster, L. (1991) 'Metalwork, wood and bone', in *The Making of England: Anglo-Saxon art and culture AD 600-900*, eds L. Webster and J. Backhouse. London: British Museum Press, 268–283.

Webster, L. (2012) *Anglo-Saxon Art*. London: British Museum Press.

Webster, L. and J. Backhouse (eds) (1991) *The Making of England: Anglo-Saxon art and culture AD 600-900*. London: British Museum Press.

Whiting, M.C. (1983) 'Appendix 2: Dye Analysis', in *The Sutton Hoo Ship-Burial volume 3: late Roman and Byzantine silver hanging-bowls, drinking vessels, cauldrons and other containers, textiles, the lyre, pottery bottles and other items*, eds R. Bruce-Mitford and A.C. Evans. London: British Museum Publications, 465.

Wickham, C. (2009) *The Inheritance of Rome: a history of Europe from 400-1000*. London: Allen Lane (repr. London: Penguin Books, 2010).

Wild, C. (1823) *An Illustration of the Architecture and Sculpture of the Cathedral Church of Worcester*. London: Charles Wild.

Wild, J.P. (1970) *Textile Manufacture in the Northern Roman Provinces*. Cambridge: Cambridge University Press.

Wild, J.P. (2003) *Textiles in Archaeology*. Princes Risborough: Shire Publications.

Williams, H. (2011) 'Mortuary Practices in Early Anglo-Saxon England', in *The Oxford Handbook of Anglo-Saxon Archaeology*, eds H. Hamerow, D.A. Hinton and S. Crawford. Oxford: Oxford University Press, 238–265.

Wincott Heckett, E. (2003) *Viking Age Headcoverings from Dublin*. Dublin: Royal Irish Academy.

Wingfield Digby, G. (1957) 'Technique and Production', in *The Bayeux Tapestry: a comprehensive survey*, ed. [Sir] F. Stenton. London: Phaidon Press, 37–55.

Wood, J. (2003) 'The Orkney Hood: an ancient recycled textile', in *Sea Change: Orkney and northern Europe in the later Iron Age AD 300-800*, eds J. Downes and A. Ritchie. Balgavies: Pinkfoot Press, 171–175.

Yeates, S. (2012) *Myth and History: ethnicity and politics in the first millennium British Isles*. Oxford: Oxbow Books.

Yorke, B. (2003) *Nunneries and the Anglo-Saxon Royal Houses*. London: Continuum.

Yorke, B. (2006) *The Conversion of Britain 600-800*. Harlow: Pearson Education.

Unpublished letters, memoranda, reports and conference papers

Anon. Unpublished memorandum, English Heritage Archive. I am grateful to Claire Tsang and Kirsty Stonell Walker, Archive and Information Team at English Heritage, Fort Cumberland for a copy of this memorandum.

de Marchi, P.M. and A.F. Palmieri-Marinoni (2014) 'Longobard Brocaded Bands from the Seprio: Production, Movement and Status', Unpublished conference paper.

Gill, K. 'Treatment Report: remounting of a group of liturgical textile fragments from Worcester Cathedral' (unpublished report), From the Textile Conservation Centre Archive. I am grateful to Christopher Guy, Worcester Cathedral Archaeologist, and Frances Lennard, Senior Lecturer in Textile Conservation, Centre for Textile Conservation, for a copy of this report.

Granger-Taylor, H. Unpublished report. I am grateful to Francis Pritchard, formerly of The Whitworth, for a copy of this work.

Pritchard, F. (2014) 'Evidence of Tablet-Weaving from Viking-Age Dublin', Unpublished conference paper.

Unpublished letter from 22 February 1957. From English Heritage Archive. I am grateful to Claire Tsang and Kirsty Stonell Walker, Archive and Information Team at English Heritage, Fort Cumberland for a copy of this letter.

WCL A402(1), Cathedral Architect's Reports, Unpublished letter from 16 June 1974. From Worcester Cathedral Archive. I am grateful to Christopher Guy, Worcester Cathedral Archaeologist, and David Morrison, Worcester Cathedral Archivist, for a copy of this letter.

Zumbuhl, M (2011) *Lexis of Cloth and Clothing Project*, Unpublished conference paper.

Websites

Acta Sanctorum Database (2012) http://acta.chawyck.co.uk (Accessed: 6 August 2012).

DRBO. ORG, 'Book of Exodus', in *Douay-Rheims Bible + Challoner Notes* (2001–2016) http://www.drbo.org/chapter/02001.htm (Accessed: 3 June 2016).

Grundy, M. (2008) 'Needlework Stitched nearly 1,000 Years Ago is Brought Back to Life', *Worcester News* http://www.worcesternews.co.uk/features/3562791.print/ (Accessed: 24 April 2010).

My Society, *Wool Textile Industry Levy*, http://www.theyworkforyou.com/debates/?id=1950-10-24a.2733.0 (Accessed: 13 October 2013).

Mumford, L. (2004) 'The Llangors Textile: an early medieval masterpiece', *WalesPast: 250,000 years of life in Wales* http://www.walespast.com/article-print.shtml?id=12 (Accessed: 4 August 2006).

Portable Antiquities Scheme, *Summary Definition of Treasury* https://finds.org.uk/treasure/advice/summary (Accessed: 13 August 2018).

Prosopography of Anglo-Saxon England (2010) http://www.pase.ac.uk/index.html (Accessed: 23 December 2011).

Schulte, A.J. (1907) 'Altar Screen', in *The Catholic Encyclopedia* http://www.newadvent.org/cathen/01356d.htm (Accessed: 22 November 2018).

The British Museum, 'Collections Online', *Trustees of the British Museum* (2017) http://www.britishmuseum.org/research/collection_online/collection_object_details.aspx?objectId=86220&partId=1&searchText=Kempston&page=2 (Accessed: 3 July 2018).

The Lexis of Cloth and Clothing Project (2014) http://lexissearch.arts.manchester.ac.uk/entry.aspx?id=3575 (Accessed: 16 January 2015).

The Textile Institute, *Textile Terms and Definitions* (2015) http://www.ttandd.org/search/ (Accessed: 21 September 2015).

University of Edinburgh, *History of the Institute* http://www.roslin.ed.ac.uk/about-roslin/history-of-the-institute/ (Accessed: 13 October 2013).

Plate 1. a) The Orkney Hood (measurements: looped stitch: 448 mm, chain stitch: c. 35 mm, 96 mm, 69 mm, 86 mm), © National Museums Scotland; b) Worcester: The embroidered fragments in their original frame, © Textile Conservation Foundation, courtesy of Worcester Cathedral.

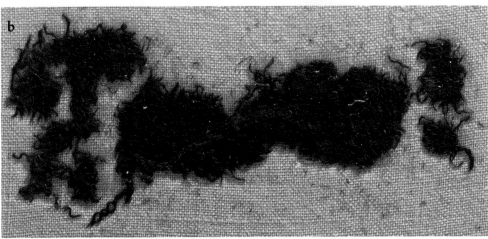

Plate 2. a) Sutton Hoo A (100 × 12–40 mm): Textile with seam covered in embroidery to the left, © Trustees of the British Museum; b) Kempston embroidery (24 × 54 mm), © Trustees of the British Museum.

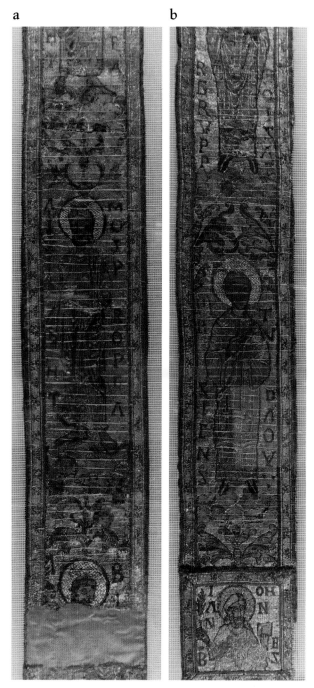

Plate 3. a) Durham D: Stole (total length 1.94 m); b) Durham E: Maniple (total measurements: 806 × 60 mm), images with the kind permission of Durham Cathedral.

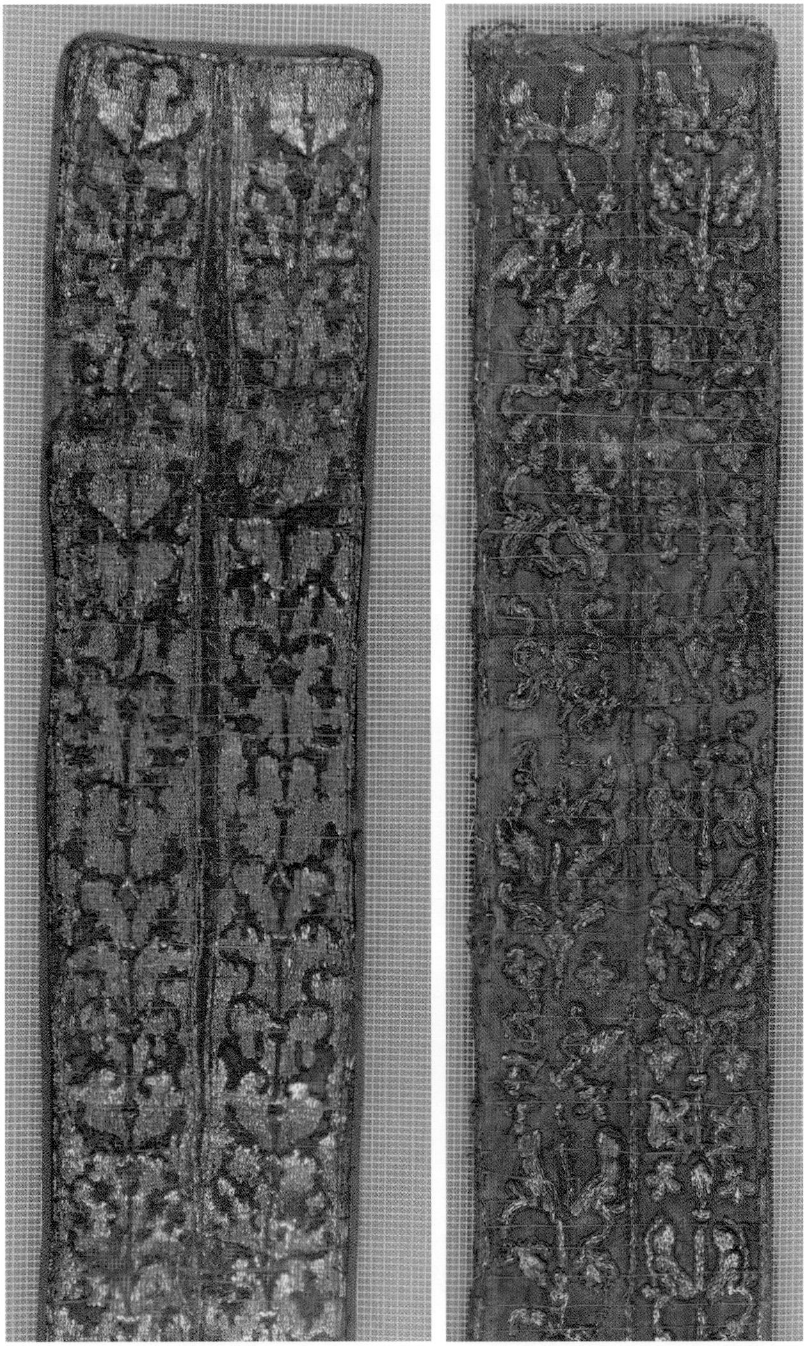

Plate 4. a) Durham C: Embroidered band known as 'Maniple II' or a girdle; a) front view (total measurements: 66 × 49 mm); b) reverse view, images with the kind permission of Durham Cathedral.

Plate 5. a) Bayeux Tapestry: Scene 30, 'Here Harold sits as King of the English' (total measurements: 68.38 × 0.5 m) 11th century, with special permission from the City of Bayeux; b) London A: shoe with embroidered top band (total measurements: 198 × 30 mm), © Museum of London.

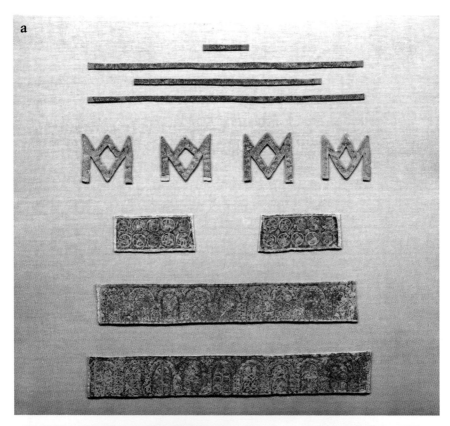
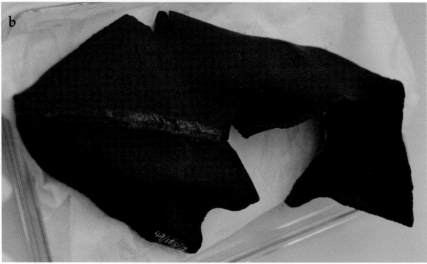

Plate 6. a) The eight Maaseik embroideries, with thanks to the Royal Institute for Cultural Heritage, Brussels; b) Coventry: Child's leather turnshoe, right foot, with embroidered vamp (embroidery: 45 × 5-7 mm), Herbert Art Gallery & Museum, Coventry.

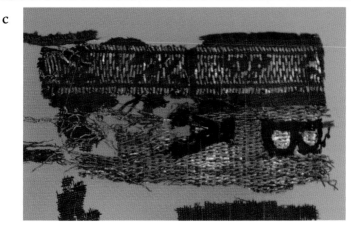

Plate 7. a–b) Ingleby: Carbonised metal thread embroidery: a) top, front view (11 × 6-7 mm); b) bottom, front view (17 × 8 mm), images Derby Museums Trust; c) Durham: Fragment of stole from Ushaw College (33 × 18 mm), with permission of the Trustees of Ushaw College.

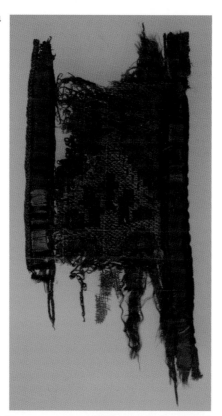

Plate 8. a) Durham A: Tablet-woven soumak band with possible embroidery (57–71 × 33 mm), with permission of the Trustees of Ushaw College; b) Durham F: Fabric with embroidery from the grave of William of St Calais (embroidery: 9–67 × 3.5–5.5 mm), with the kind permission of Durham Cathedral.

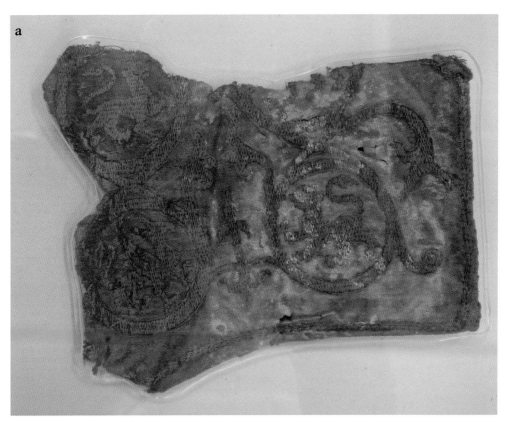

Plate 9. a) Durham G: Fabric with embroidery from the grave of William of St Calais (embroidery: 36–71 × 16–92 mm), with the kind permission of Durham Cathedral; b) Milan: Gold work embroidery in its 1940s frame (embroidery: 345–347 × 29–32 mm), with thanks, Archivo e biblioteca della Basilica di S. Ambrogio.

Plate 10. The composite textile or casula of Sts Harlindis and Relindis, with the eight embroideries in situ, with thanks to the Royal Institute for Cultural Heritage, Brussels.

Plate 11. Oseberg J: Colour drawing of section 12B5 (145 × 80 mm), © Museum of Cultural History, University of Oslo, Norway.

Plate 12. a) Milan: Detail of gold embroidery attached to the original silk ground fabric (×210), with thanks, Archivo e biblioteca della Basilica di S. Ambrogio; b) Durham A: Detail of faded coloured threads (×400); c) Durham A: Detail of possible embroidered wrapping stitch (×400), images with permission of the Trustees of Ushaw College; d) Kempston: copper alloy box in which the embroidery was found, © Trustees of the British Museum.

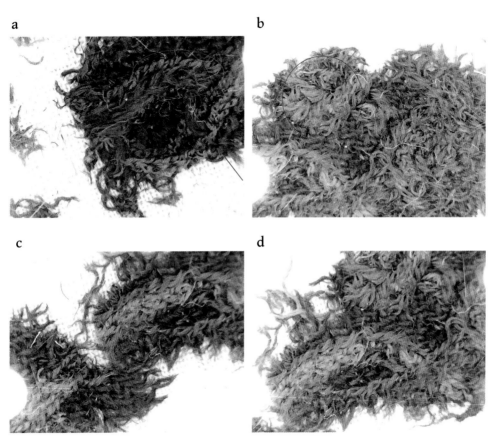

Plate 13. Microscopic images of the Kempston embroidery (×400); a) chain stitch circled. The arrow to the right points to the possible split stitch; b) line of stem stitch circled; c) the two sections placed next to each other; d) the break in stitching and the small individual stitches, images © Alexandra Makin (2013), taken courtesy of the Trustees of the British Museum.

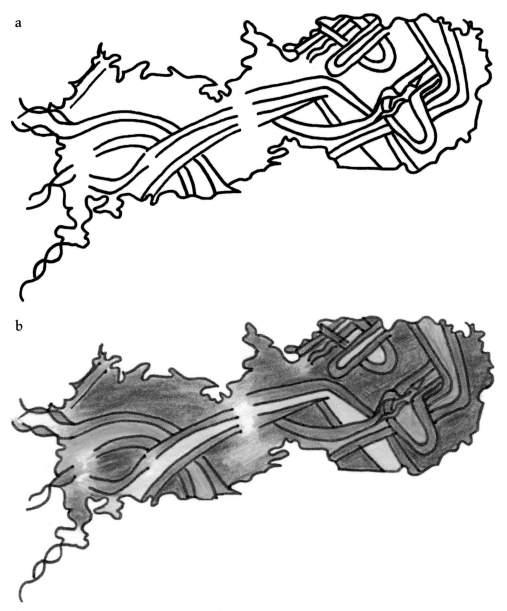

Plate 14. Kempston: a) line drawing and b) coloured drawing of embroidery design, © Alexandra Lester-Makin.

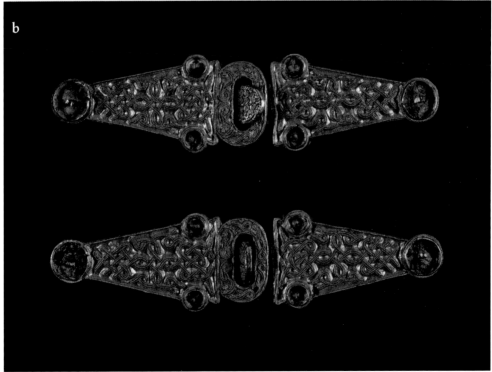

Plate 15. a) Great gold buckle, burial mound 1, the ship burial, Sutton Hoo; b) shoulder clasps, Taplow, images © Trustees of the British Museum.

Plate 16. Book of Durrow carpet page, MS57 folio 192v, Board of Trinity College Dublin.

Plate 17. a) Kingston broach: detail, courtesy National Museums Liverpool: World Museum; b) Sword hilt collar from the Staffordshire Hoard, © Birmingham Museums Trust.

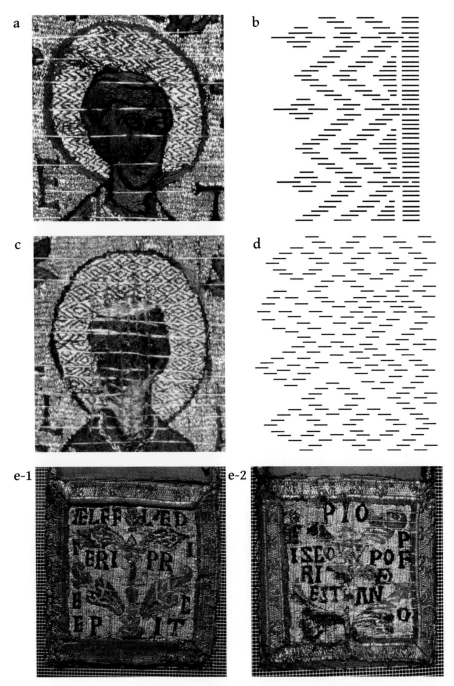

Plate 18. a) Durham E: maniple, detail, halo of the deacon Peter; b) line drawing, braid 9 from Durham; c) Durham E: maniple, detail, halo of Pope Gregory; d) line drawing, braid 10 from Durham; e) Durham D: Stole both end tabs, reverse, images a, c, e with the kind permission of Durham Cathedral; b, d © Alexandra Lester-Makin, after Crowfoot (1956).

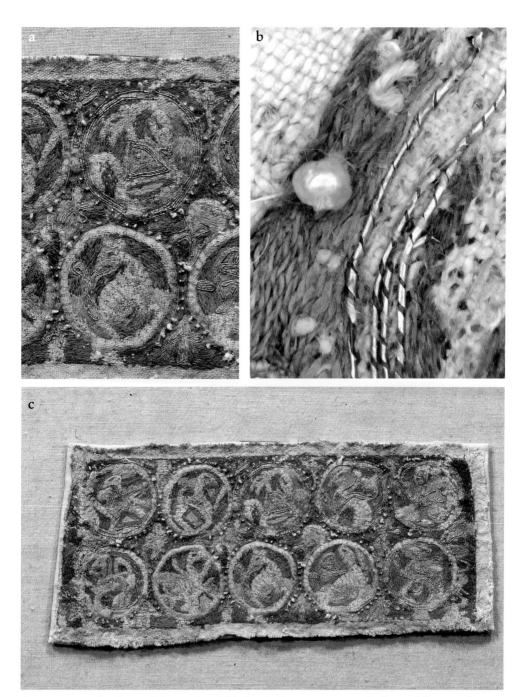

Plate 19. Maaseik: a) roundel strip detail; b) detail of a surviving pearl; c) roundel strip 2 (c. 190 × 187 mm), images a and c with thanks to the Royal Institute for Cultural Heritage, Brussels, b with thanks to the Tourism Board of Maaseik and the Royal Institute for Cultural Heritage, Brussels.

Plate 20. a) Books of Kells: moths highlighted, MS58 folio 34r, Board of Trinity College Dublin; b) Milan: detail of couched gold thread (×400), with thanks, Archivo e biblioteca della Basilica di S. Ambrogio; c) Durham D: stole, detail of couched gold thread (×400), with the kind permission of Durham Cathedral; d) Maaseik B: detail of couched gold thread (×400), with thanks to the Tourism Board of Maaseik and the Royal Institute for Cultural Heritage, Brussels.

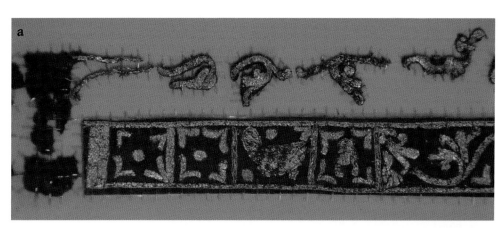

Plate 21. Milan a) second square from left: duck with head turned back, third square from left: stylised insect; b) detail of gold work design and layout, images with thanks, Archivo e biblioteca della Basilica di S. Ambrogio.

Plate 22. Dedication page of Bede's Life of St Cuthbert, MS183 folio 1v, Parker Library, Corpus Christi College, Cambridge.

Plate 23. Maaseik A: arcade strip detail, with thanks to the Royal Institute for Cultural Heritage, Brussels.

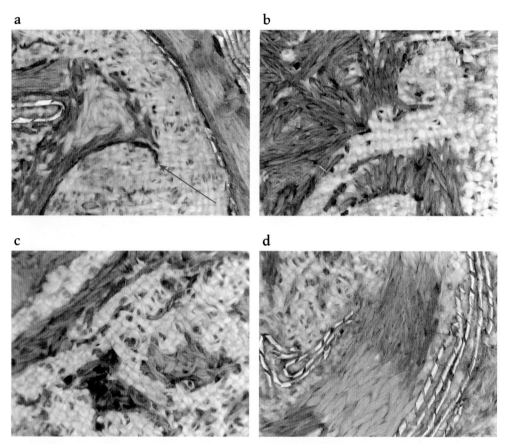

Plate 24. Maaseik A (×400): a) detail, outlines worked first then reworked after the filling is complete (highlighted by arrow); b) detail, stitching following triangular motif shapes; c) detail, more randomly orientated stitching; d) detail, stitching follows line of curve and colours are worked in blocks, images with thanks to the Tourism Board of Maaseik and the Royal Institute for Cultural Heritage, Brussels.

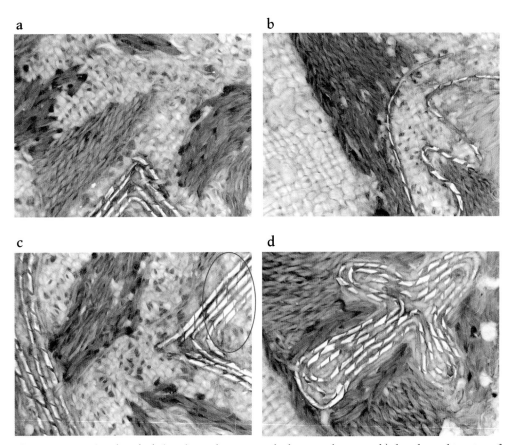

Plate 25. Maaseik B (×400) a) detail, stitches are worked in one direction; b) detail, combination of stitched lines and chevrons; c) detail, couching stitches worked in rows (circled); d) detail, couched gold thread, images with thanks to the Tourism Board of Maaseik and the Royal Institute for Cultural Heritage, Brussels.

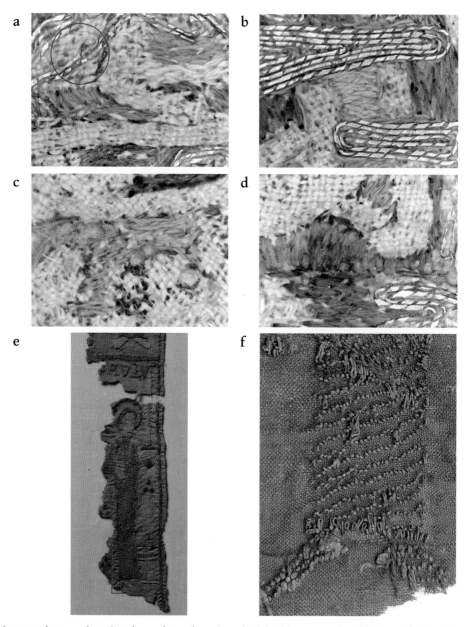

Plate 26. a) Maaseik A: detail, couching thread worked double over each gold thread (circled) (×400); b) Maaseik B: detail, gold thread bent precisely round corners (×400); c) Maaseik A: detail, filling threads cross outline threads (×400); d) couched gold catches silk infilling threads (×400), images with thanks to the Tourism Board of Maaseik and the Royal Institute for Cultural Heritage, Brussels; e) and f) Worcester: details of underside couching used as a filling stitch, front and reverse, images reverse © Textile Conservation Foundation, courtesy of Worcester Cathedral.

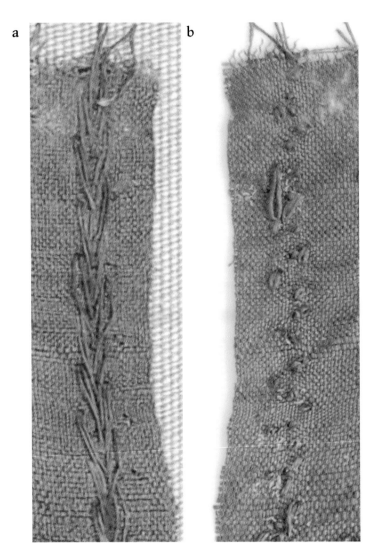

Plate 27. Worcester: a) detail of underside couching used as an outline, front; b) detail of underside couching used as an outline, reverse images © Textile Conservation Foundation, courtesy of Worcester Cathedral.

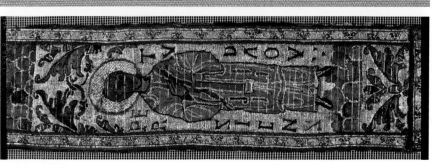

Plate 28. Comparative examples of silk work: a) Durham E, with the kind permission of Durham Cathedral; b) Worcester, © Textile Conservation Foundation, courtesy of Worcester Cathedral; c) orphrey (AD 1310–1325), museum number: T.72-1922, © Victoria and Albert Museum, London.

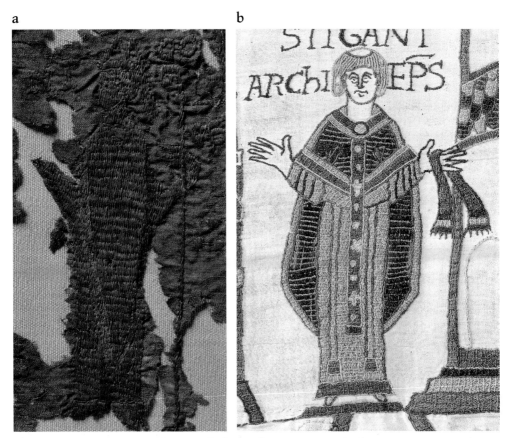

Plate 29. Comparative stance between bishops: a) Worcester, © Textile Conservation Foundation, courtesy of Worcester Cathedral; b) Bayeux Tapestry, detail, 11th century, with special permission from the City of Bayeux.

Plate 30. Comparative stance between figures: a) Worcester, © Textile Conservation Foundation, courtesy of Worcester Cathedral; b) Bayeux Tapestry, detail, 11th century, with special permission from the City of Bayeux.

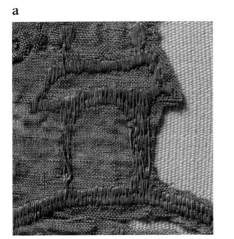

Plate 31. Comparative buildings: a) and c) Worcester, © Textile Conservation Foundation, courtesy of Worcester Cathedral; b) and d) Bayeux Tapestry, detail, 11th century, with special permission from the City of Bayeux.

Plate 32. Maaseik C: two of the four monograms (120 × 120/131 mm), with thanks to the Royal Institute for Cultural Heritage, Brussels.